Electric Power System Fundamentals

REVISED AND EXPANDED **SECOND EDITION**

by
Robert M. Clough

Electric Power System Fundamentals

Copyright © 2011, 2013 by Robert M. Clough

Published by Clay Bridges Press in Brenham, TX.

All rights reserved. No part of this publication may be reproduced, stored in a retrieval system, or transmitted in any form by any means, electronic, mechanical, photocopy, recording, or otherwise, without the prior permission of the publisher, except as provided for by USA copyright law.

First Printing 2013

ISBN-13: 978-1-939815-06-4

ISBN-10: 1939815061

Special Sales: Most Clay Bridges Press titles are available in special quantity discounts. Custom imprinting or excerpting can also be done to fit special needs. Contact Clay Bridges Press at marissa@claybridgespress.com.

Photo and Illustration Credits:

Robert M. Clough - pages 9, 12, 13, 19, 20, 22, 26, 27, 30, 34, 54, 56, 57, 58, 59, 60, 65, 68, 75, 79, 80, 81, 83, 85, 86, 90, 92, 94, 95, 96, 98, 99, 107, 108, 110, 112, 113, 117, 119, 121, 122, 133, 138, 144, 145

Stewart & Stevenson, LLC - pages 16, 23, 29, 32, 38, 40, 63, 69

Page 24 (The Right Hand Rule): author: Jfmelero

Page 87: *(WIKIPEDIA COMMONS) picture 'retrieved from* "http://commons.wikimedia.org/w/index.php?title=File:Leistungsschalter-110KV.jpg&oldid=70684023"'

Page 88: *(WIKIPEDIA COMMONS) 'retrieved from* "http://en.wikipedia.org/wiki/File:CurrentTransformers.jpg"' - author: ali@gwc.org.uk

Page 89: *(WIKIPEDIA COMMONS) Voltage Current Transformers - picture 'retrieved from* "http://en.wikipedia.org/wiki/File:SF6_current_transformer_TGFM-110_Russia.jpg"'

The author assumes no liability with respect to the use of the information contained herein.

Preface

This book presents a beginning study of electric power systems from generation and control to the analysis of power grid networks. In treating such a broad topic, detailed discussions of subjects such as generator synchronism, protective relaying circuitry and battery chemistry have been relegated to appendices rather than allowed to divert the central discussion into areas better suited for later detailed study. As a beginning study, only elementary algebra and trigonometry are used and discussions are based upon intuitive rather than rigorous mathematical proofs. The book is intended to assist individuals in acquiring a conceptual background in electric power systems whether to advance their careers or to determine in what part of the industry they may wish to specialize.

Questions are included after major sections of the book for use as a learning aid. Projectable images of illustrations from the book for classroom use and additional book copies are available through epsf@sbcglobal.net.

Electric Power System Fundamentals

In creating this book I am particularly indebted to my long term mentor, Roy T. Allice of Stewart & Stevenson, LLC, to John C. Pierce, Program Chair for Energy Generation Operations at Southeast Community College in Milford, Nebraska, to Charles A. Heller, Ph.D. of Blinn College for his assistance in matters related to chemistry, to Jerry W. Lester, Ph.D. a former research scientist at The University of Texas Medical Branch who patiently read the early manuscript and whose criticisms were immensely helpful in improving it, and to Janet Gonzales for her editorial review and commentary. I must also thank my loving wife, JoAnn for her constant support.

Robert M. Clough
Chappell Hill, Texas
September 2013

Table of Contents

1 Introduction 1

2 Electric Power Generation 9

    2.1 Generator Construction 14

    2.2 Generator Controls 45

        2.2.1 Regulating the Prime Mover Throttle 45

            2.2.1.1 Electric Power Definition .. 45

            2.2.1.2 Mechanical and Electrical Power 48

        2.2.2 Regulating Rotor Excitation Current 50

            2.2.2.1 Inductive Reactance 51

            2.2.2.2 Capacitive Reactance 52

            2.2.2.3 Resistance 53

            2.2.2.4 Impedance 53

        2.2.3 Control Effects on Generated Power 56

        2.2.4 Generator Safe Operational Boundaries 62

		2.2.5	Control Implementation and Operation 64
			2.2.5.1 Isochronous Control 66
			2.2.5.2 Parallel Control 67

3 Electric Power Transmission And Distribution 73
    3.1 Transformers 75
    3.2 Transmission and Distribution Wiring 91
    3.3 Substation Functions and Controls 95

4 Electric Power Grid Management 105
    4.1 Grid Power Flow 105
    4.2 Contingency Analysis 115
    4.3 Load Prediction 118
    4.4 Economic Operational Considerations 120
        4.4.1 Economic Dispatch 121
        4.4.2 Unit Commitment 123

5 Appendices 127
    A. Battery Chemistry 129
    B. Generator Synchronization 137
    C. Generator Protective Circuitry 143
    D. Glossary of Terms 157

6 Answers to Questions 183

7 Bibliography 193

1

INTRODUCTION

For thousands of years, humankind with all of its achievements had only animals, windmills, and waterwheels to lighten its physical loads and fire to light its dwellings. Then within the span of a mere 100 years a chain of discoveries, each building upon the other, led to the world-wide use of electric power with all its benefits. These discoveries began in Bologna, Italy in 1791 and ended in 1893 at the Chicago World's Fair. It is fair to say *ended* because there have been no significant changes in the fundamental methods of generating or distributing electric power in the succeeding 120 years.

The discovery in 1791 that initiated the evolution of modern electric power systems is credited to an Italian scientist, Luigi Galvani. Galvani's name is carried to us today in the words Galvanize and Galvanometer. In 1791 he was a teacher at the University of Bologna and classified himself a physiologist. He is perhaps most noted for laying the foundation material in the study of neurology. As a physiologist he experimented with applying static (instantaneous) electric charges to human limbs that were paralyzed or immobile. Static electric charges were created by rubbing cloth and glass together. In more exacting studies of muscular reactions under laboratory conditions, Galvani dissected frog's legs and gauged their reactions to electric charges. In this process, he discovered frog muscles reacted when their ends were connected between metals with no charge applied. He recorded his findings in the scientific journals of the day and explained the phenomena resulted from "animal electricity," created within the frog's legs themselves.

Ten years later another Italian scientist, Alessandro Volta, for whom the unit of electric potential the Volt would later be named, began duplicating Galvani's experiments. Volta discovered the frog muscles reacted only when the metals that connected their ends were *different* metals such as zinc and copper and postulated the frog legs were reacting to an electric current flowing between the two dissimilar metals. Volta then replaced the frog legs separating the two metals with layers of cardboard saturated with a solution of salt and proved that a current was generated between the metals. Volta, realizing the significance of this primitive demonstration, soon developed methods of increasing the amount of current produced by stacking successive layers of dissimilar metal separated by what became known as "electrolytes." These stacks became known as "Voltaic Piles" and were in fact the world's first batteries. The scientific community now had a reliable and continuous source of electricity for experimentation. It was well known that a conductor connected between the ends of a Voltaic Pile created heat from a new and mysterious source.

In 1820, almost 30 years after Galvani wrote of his

Introduction

frog leg experiments and 20 years after Volta invented his Voltaic Pile, a physicist named Hans Oersted at the University of Denmark made an accidental discovery. While demonstrating the properties of a Voltaic Pile to his students, Oersted noticed the needle of a compass lying nearby deflected away from pointing north when the ends of the Voltaic Pile were connected but returned to point north when the connection was opened. Experimenting further, Oersted found the compass deflected in the opposite direction when exposed to the opposite direction of current flow. His conclusion that swept through the scientific community was that an invisible field linked electricity and the magnetic compass affecting it in the same way as if a magnet had been placed near the compass. A link, therefore, between electricity and magnetism existed raising the question: If an electric current could produce a magnetic field, could a magnetic field produce an electric current?

The answer to the question was demonstrated eleven years later in 1831 by Michael Faraday, a scientist at the Royal Institution in London. Faraday's demonstrations conclusively proved that a current is produced in a conductor when the conductor is moved through a magnetic field. Without motion, however, no current was produced. There are several corollaries to this finding that will be discussed later in this text but the generation of electricity in today's electric power systems is due to metallic conductors moving through magnetic fields.

Following Faraday's demonstrations, a great many electric generators were built in various configurations designed to move electric conductors through magnetic fields. Because the magnetic fields and conductors were

Electric Power System Fundamentals

required to be in motion, the currents produced alternated with the motion. The alternating current phenomenon challenged designers in how to build electric motors that rotated continuously rather than oscillated back and forth with the alternating current. As a consequence, the alternating cycle of electric current was taken from rotating generators through what were termed, "split commutators." Using this mechanism the alternating current was converted to flow in one direction only and was called "direct current" rather than "alternating current."

The well known American inventor, Thomas Edison, using direct current generators, built the first commercial electric power plant in the United States on Pearl Street in New York City in 1882. His purpose was to sell his electric light bulbs to a wider market, billing the power he furnished to customers based upon the number of light bulbs in their respective homes and offices. The direct current he produced, however, lost its potential after being transmitted only a few city blocks from the generating station. To overcome the difficulty, generating plants would be required at intervals of every few blocks to electrify New York City and an uncountable number would be needed to electrify the entire country.

At about this time, a Serbian electrical engineer named Nicola Tesla, arrived in New York and began working for Edison. His story is interestingly told in the book: *Tesla: Man out of Time*, by Margaret Cheney (See Bibliography). Tesla and Edison could not have been more different in their life styles or in their approaches to science. Tesla was neat in appearance, while Edison seldom bathed. Edison continuously hammered on problems, trying one solution after another until a

Introduction

solution was finally found. Tesla, on the other hand, never picked up a tool to work on anything until he had the solution worked-out in his mind. While strolling in a Budapest park in Hungary, Tesla created in his mind a conceptual alternating current power system including electric motors that would run on alternating current. The overall concept was revolutionary but very efficient and practical in every aspect. It also included transformers that had been invented by Michael Faraday to elevate generated voltages reducing losses in long distance power transmission.

Tesla needed a machine shop and laboratory to build working models of his alternating current equipment but, Edison was committed to the use of direct current and refused to listen or invest in an alternating current system.

Tesla then took his ideas to the industrialist and well known American engineer, George Westinghouse. Westinghouse had invented the air-driven braking system for steam trains and was involved in building heavy machinery. He immediately understood the practical potential of Tesla's alternating current electric power system and formed a partnership with Tesla. With the new found funding, Tesla constructed working models of the AC system including alternating current generators, motors, and transformers. The practicality and measured performance of the alternating current system presented an obvious advantage over existing direct current systems and after it was successfully demonstrated at the Chicago World's Fair in 1893, the Tesla-Westinghouse alternating current system won the contract to build the first two generators at Niagara Falls in October 1893. As of the end of the first decade of the 21st century, there have been no significant changes in the electric power system first presented by Nicola Tesla to Thomas Edison in the 1880s. Gardner H. Dales of the Niagara Mohawk Power Corporation addressing the Institute of Radio Engineers, later to become the American Institute of Electronic and Electrical Engineers (AIEE), on April 5, 1956 said:

> "If ever there was a man who created so much and whose praises were sung so little—it was Nikola Tesla. It was his invention, the poly-phase [alternating current] system, and its first use by the Niagara Falls Power Company that laid the foundation for the power system used in this country and throughout the entire world today..."

The following chapters discuss the details of these discoveries and how they were implemented to construct and control the vast networks that provide electric power to millions of users throughout the world.

Introduction

Section 1 Questions:

1. What important question was answered in Faraday's demonstrations?

2. Why did Volta need Galvani's experimental evidence to create his invention?

3. Why did Oersted need Volta's invention to make his discovery?

4. What important question resulted from Oersted's observations?

5. What caused Tesla's system of poly-phase AC to become more widely used than Edison's DC power system?

6. What was the single most important aspect of Tesla's AC system and whose discovery first demonstrated the phenomenon?

7. What was the name of Volta's invention and what was it later called?

Notes

2

ELECTRIC POWER GENERATION

Figure 2.0
Typical Coal Fired Power Plant

Electric power is generated by converting heat into electricity in facilities such as the coal burning power

Electric Power System Fundamentals

plant illustrated in Figure 2.0 (◄). In this plant, heat from burning coal converts water into steam in boilers. The steam is then directed through turbines that convert the flow of expanding steam into rotational shaft horsepower. Solar heat produces wind by differential heating of the Earth's surface to drive windmills. Solar heat also lifts humid air to atmospheric levels where it cools and condenses into rain and snow, filling streams and lakes that drive waterwheels. Regardless of the mechanism used to do so, power plants rotate the shafts of electric generators from converted heat.

Electric utility companies need to know the cost of converting heat into electric power in calculating power production costs and to determine when, how long, and at what load levels individual power plants should be brought on and off line. Power plant efficiency also relates to when plants should be shut-down for maintenance and even when re-design or expansion projects should be considered. The efficiency with which a power plant converts fuel-to-heat-to-horsepower-to-watts of electrical energy is defined as its heat rate. Therefore, a plant's overall efficiency may be expressed as:

$$= \frac{\text{Heat Content of the Fuel Used}}{\text{Generated Power Output (Watts)} - \text{Parasitic Load}}$$

$$= \frac{\text{Btu*}}{\text{kWh**}}$$

* Heat required to increase the temperature of 1 lb. of water 1°F.

** Includes parasitic load

Electric Power Generation

Parasitic load is the power used within a power plant to operate fuel pumps, lubrication and water circulation pumps, to provide power for utility lighting and control systems, and other miscellaneous equipment.

The heat content of a particular fuel is determined from measurements in a laboratory in Btu per unit weight or unit volume.

Fuel Type	Heat Value*
Coal	25×10^6 Btu/ton
Crude Oil	5.6×10^6 Btu/bbl
Gasoline	5.2×10^6 Btu/bbl
Natural Gas	1,030 Btu/ft^3

* Nominal values; measured values may differ

Table 2.1
Fuel Types and Heat Content

Table 2.1 (▲) lists the fossil fuels most commonly used in electric power plants and their respective heat contents per unit weight or volume.

The heat rate of nuclear fuel is not included in Table 2.1 and is meaningless when considering the weight of fissionable material in a reactor compared to its potential heat release. Nuclear energy is created when mass is destroyed. Energy and mass are expressed in Einstein's well-known equation:

$$E = mc^2$$

The velocity of light (c) is 186,000 miles per second; therefore, a very small change in mass (m) will produce an enormous amount of energy. For one pound of mass, the potential energy released is:

$$\frac{E}{lb} = 38,690 \underline{\text{billion}} \text{ Btu}$$

Converting all of the mass to energy, however, is not yet possible. Only about 1×10^{-3} of the mass is consumed in a typical uranium reactor per pound of uranium or about 3 million times more energy than is obtained by burning one pound of coal.

The following example illustrates how the heat rate of a combustion gas turbine* running on natural gas is found:

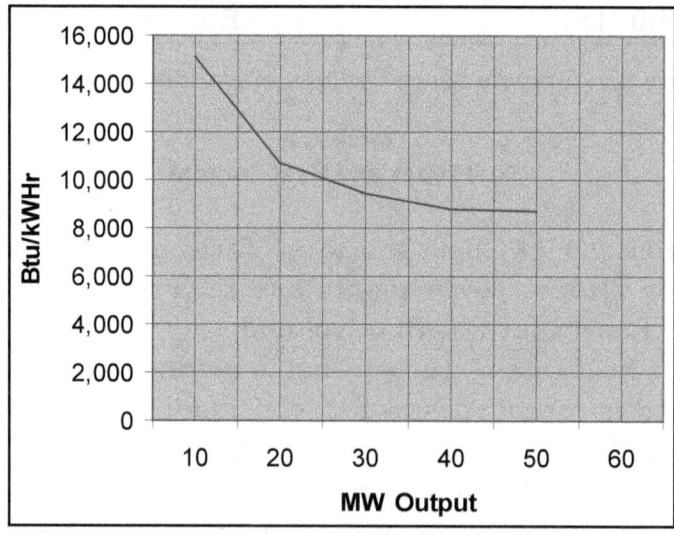

Figure 2.2
Combustion Gas Turbine Running on
Natural Gas Heat Rate vs. Output Load

---

* Combustion gas turbines convert expanding fuel-heated air into shaft horsepower.

Electric Power Generation

Figure 2.2 (◄) was derived from measured fuel flow rates vs. measured generator output in megawatts*. At the 25 MW output point, the table indicates 10,000 Btu per kilowatt†-hour were being consumed.

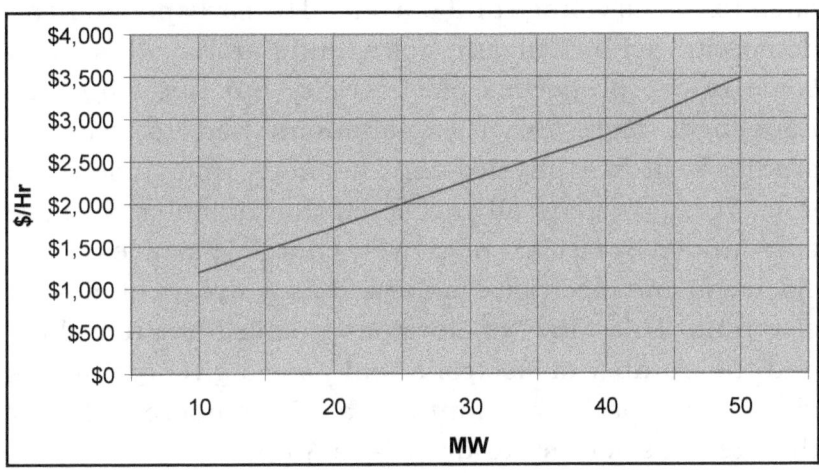

Figure 2.3
Combustion Gas Turbine Running on
Natural Gas Cost per Hour vs. Output Load

Figure 2.3 (▲) is a cost per hour graph derived by using a natural gas price of $8 Per 10^6 Btu as follows:

$$\frac{\$}{\cancel{10^6 \text{ Btu}}} \times \frac{\cancel{10^6 \text{ Btu}}}{\cancel{\text{MW}} \text{ Hr}} \times \cancel{\text{MW}} = \$/\text{Hr}$$

For example, the $2,000 per hour cost at the 25 MW point is found as follows:

$$\frac{\$8 \times 10}{\cancel{10^6 \text{ Btu}}} \times \frac{\cancel{10^6 \text{ Btu}}}{\cancel{\text{MW}} \text{ Hr}} \times 25 \cancel{\text{ MW}} = \$2000/\text{Hr}$$

---

* 1 megawatt = 1 × 10^6 Watts
† 1 kilowatt = 1 × 10^3 Watts

Electric Power System Fundamentals

2.1 GENERATOR CONSTRUCTION

A discussion of generating electricity requires a definition of what electricity is and what is meant when we use such terms as voltage and current. The early phenomena demonstrated by Galvani, Volta, and Faraday illustrated some of the properties of electricity but not *why* they happened. More exact explanations had to wait for studies in atomic physics and chemistry that came years later. It is now generally accepted that atomic nuclei are surrounded by circling negatively charged particles called electrons and electricity is defined as a stream of moving electrons. This "flow" of electrons is called "current flow" and the number of electrons that pass a given point in a given time has been quantified as the ampere to honor the French scientist, Andre-Marie Ampere.

As electrons move through materials that conduct electricity, a deficiency of electrons is caused at the point from which electrons are taken with a corresponding surplus at the point at which they are received. The point with the surplus of electrons is said to be negatively charged and the point with the deficiency is said to be positively charged. Units of charge difference between charged points are quantified in "Volts."

The forces that cause electron flow are called electromotive forces (emf). Volta's zinc, copper, and saltwater "battery" created an electro-motive force that simulated the charge difference across the ends of the frog's legs in Galvani's experiments; however, the batteries that soon emerged from Volta's laboratory using different materials suggest that the first battery merely demonstrated a principal that motivated him to create batteries for more

Electric Power Generation

practical uses. The chemical reactions that form the basis for modern batteries are discussed in Appendix A.

A convention was established before these phenomena were fully understood, stating that current flows from positively charged points to negatively charged points. This convention, although incorrect, is referred to as the direction of *conventional* current flow, and unless otherwise noted, will be used throughout this text.

Michael Faraday demonstrated a mechanical method of creating an emf without chemical cells. It should be emphasized that Faraday was merely attempting to answer the question posed by Oersted's experiments with a battery and a magnetic compass, i.e., "If an electric current could produce a magnetic field, could a magnetic field create an electric current?"

Electric Power System Fundamentals

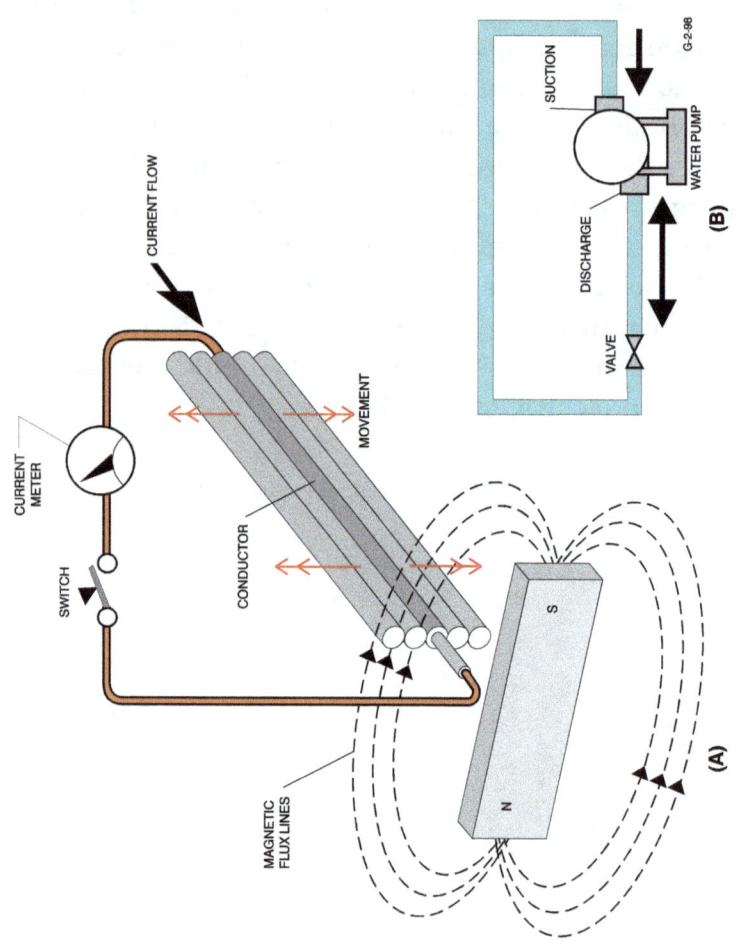

**Figure 2.1-1
Michael Faraday's Demonstration**

Electric Power Generation

Faraday's demonstration is illustrated by the arrangement of a magnet and a conducting loop of wire in Figure 2.1-1(A) (◄).

The simple demonstration revealed several facts:

1. The current measuring meter in the electrical conductor circuit will indicate current flow in the conductor only when there is relative motion between the permanent magnet and the conductor.

2. When relative motion between the magnet and conductor is reversed, i.e., toward or away from each other, the direction of current flow also reverses,

3. The magnitude or amplitude of the current produced by relative motion is directly proportional to the rate of change of motion, and

4. Current flow in the conductor is directly proportional to the field strength of the magnet, e.g., a stronger magnet produces more current at any given rate of motion than one of lower field strength.

At the atomic level, the emf created within the conductor is caused by electrons in the conductor being repelled by their reaction to magnetic forces in one direction and attracted by magnetic forces in the opposite direction. The electron charges are thereby propelled to move within the conductor depending upon the direction in which the magnetic field is moving them.

Figure 2.1-1(B) (◄) illustrates a mechanical analogy in which fluid in a pipe connected between the suction and discharge ports of a mechanical pump moves when the pump shaft is rotated just as current does in the electrical conductor when relative motion occurs between the magnet and conductor. In either case, mechanical work must be exerted to move liquid or electric current.

When the valve in the pump liquid circuit is closed, the pressure difference between the suction and discharge ports remains present just as the voltage difference remains when the switch is open in the electrical circuit. In both cases the flow of liquid or current is stopped.

Figure 2.1-2 (►) illustrates a conceptual generator that produces alternating current by rotating a permanent magnet adjacent to a fixed loop conductor. The direction of flux lines out of the North Pole and into the South Pole of the permanent magnet remain the same as the magnet is rotated by the hand crank, reversing the direction that flux lines pass the stationary conductor during each cycle of rotation. The maximum flux density at each pole maximizes the emf generated as the north and south poles alternately pass the conductor. The generated emf passes through zero as the flux lines reach minimum density and changes polarity as the opposite pole approaches.

Electric Power Generation

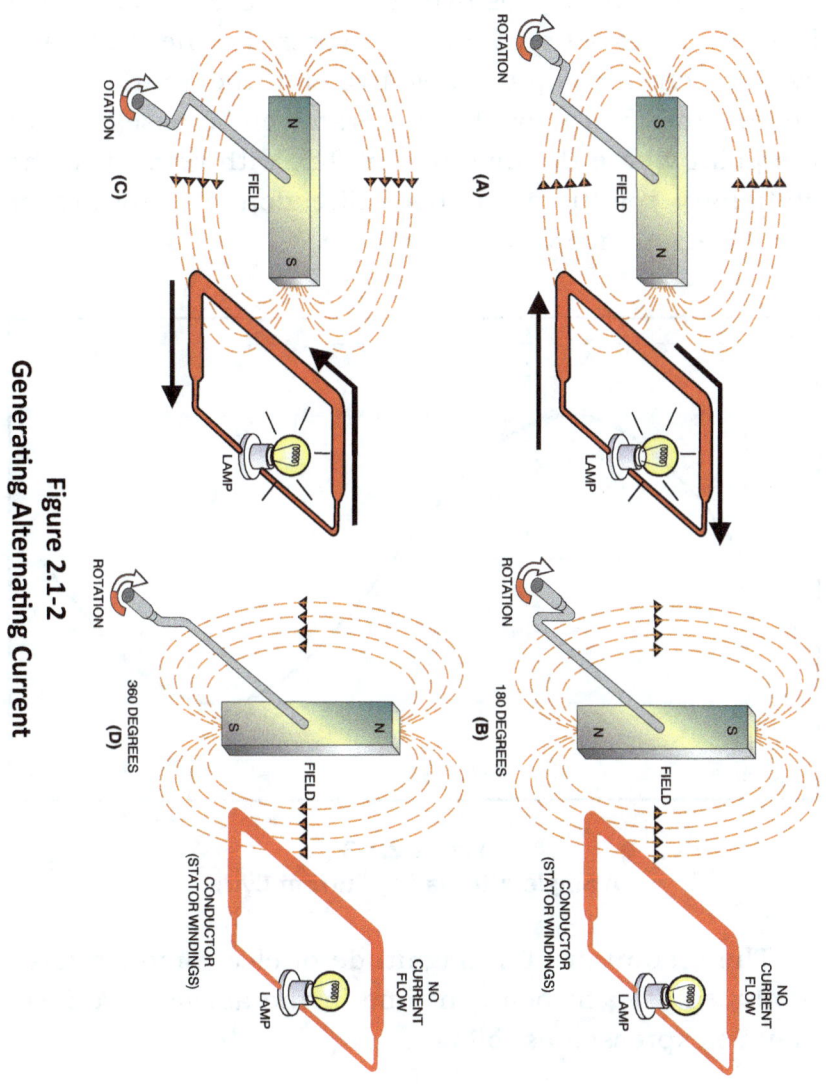

**Figure 2.1-2
Generating Alternating Current**

19

A constant speed of rotation will produce a sine wave emf wave-form as illustrated in Figure 2.1-3 (▼). At the maximum positive emf half cycle, the North Pole is nearest the conductor. When the negative half cycle reaches maximum negative emf, the South Pole is nearest the conductor. The generated emf crosses zero at the point where the direction of flux with respect to the stationary conductor changes direction, i.e., toward or away from or vice versa.

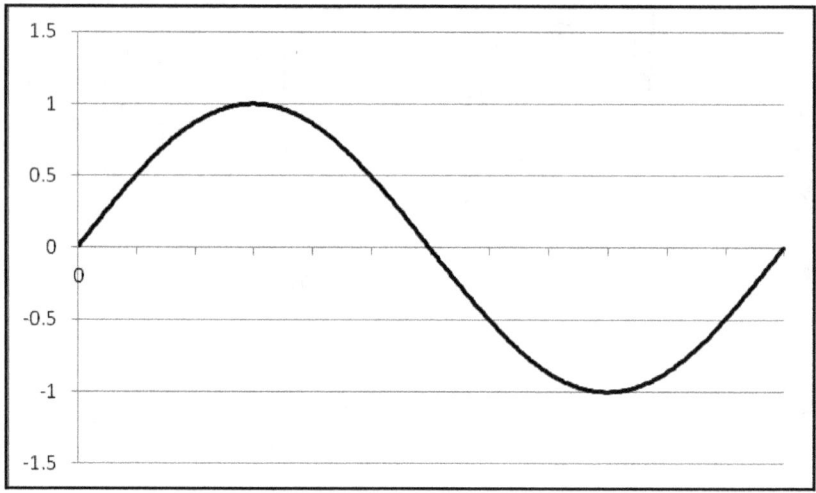

**Figure 2.1-3
A Single Alternating Current Cycle**

The instantaneous magnitude of electromotive force produced at each point in time in the above wave form may be expressed as follows:

$$e = \Phi \, (d\Phi/dt)$$

Electric Power Generation

where,

> e = instantaneous electro motive force that is generated
>
> Φ = magnetic field strength
>
> $(d\Phi/dt)$ = rate of change of the magnetic field with respect to time

In a generator with a single rotating pole-pair, as illustrated in Figure 2.1-2, one revolution of the rotor shaft produces one cycle of output voltage. To avoid future confusion, such a generator is called a single pole generator. The standard rate of change of alternating current in the United States is 60 cycles per second, although in much of the world the standard is 50 cycles per second. To generate 60 cycles per second the rotor must rotate at 60 revolutions per second or 60 cps × 60 sec/min = 3600 revolutions per minute. In generating 50 cycles per second, the rotor speed is 50 cps × 60 sec/min = 3000 revolutions per minute. In generators that must run at slower speeds such as in wind turbines or hydro-electric turbines, more poles are added to the rotor to lower the generator speed required. Each added pole pair, halves the speed needed.

Nicoli Tesla visualized that if three conductors were used on the stationary or static part of the machine rather than one, three mutually exclusive outputs could be generated per revolution. Figure 2.1-4 (▶) illustrates this concept commonly called three-phase generation. Spacing the conductors at 120 mechanical degree intervals creates even spacing between their respective output wave forms as illustrated by the three colors in Figure 2.1-5 (▶).

Electric Power System Fundamentals

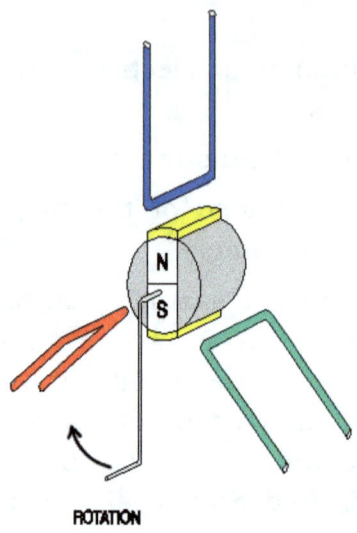

**Figure 2.1-4
Tesla's Invention**

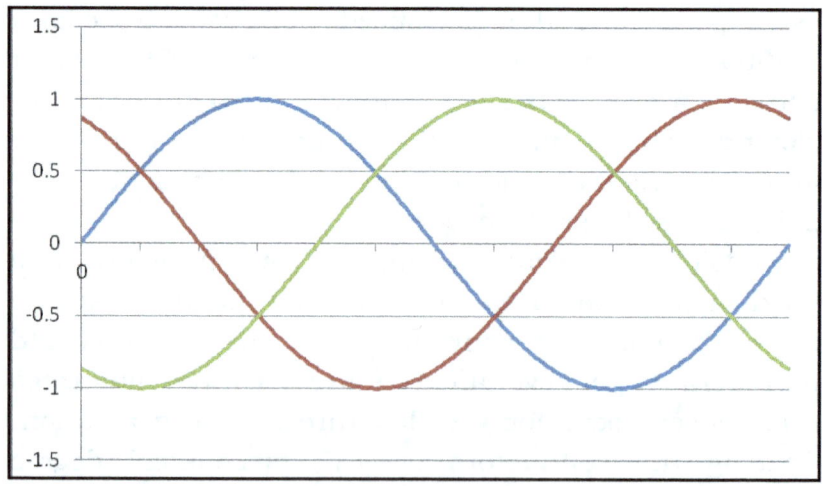

**Figure 2.1-5
Three Phase Generation**

Electric Power Generation

Generators in commercial power plants use coils of wire to form the rotor and stationary elements rather than the rotating permanent magnet and single stationary conductors as illustrated in Figure 2.1-4 (◄).

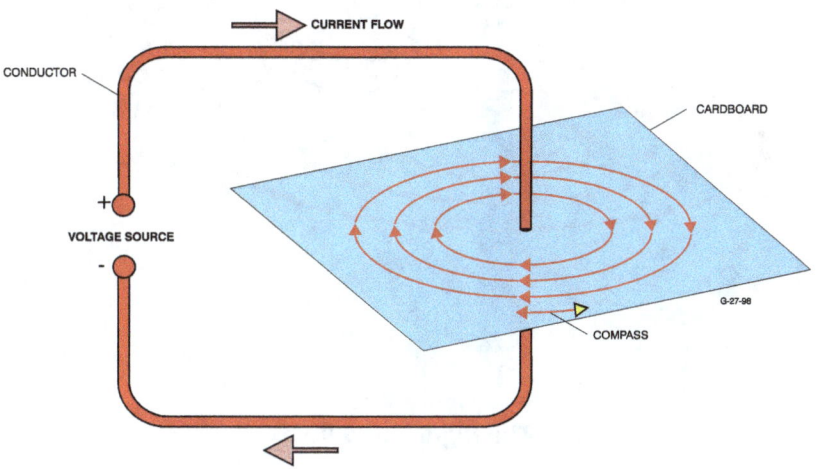

**Figure 2.1-6
Oersted's Discovery**

Figure 2.1-6 (▲) illustrates Oersted's discovery relating electricity and magnetism. Current flow through a conductor caused a compass to deflect away from pointing north and align with the invisible lines of magnetic flux surrounding the conductor.

The direction of the magnetic flux lines is predictable by the "right-hand-rule" illustrated in Figure 2.1-7 (►).

Electric Power System Fundamentals

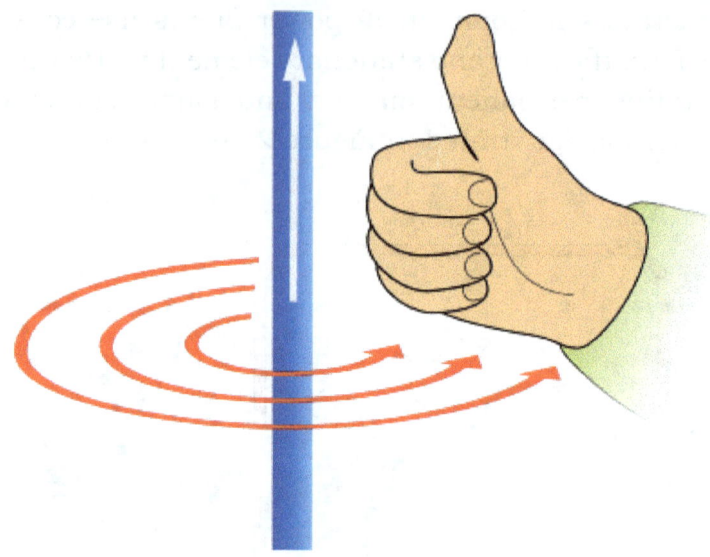

**Figure 2.1.7
The Right-Hand Rule**

The curl of the fingers indicates the direction of flux lines around a conductor and the thumb points in the direction of the current flow.

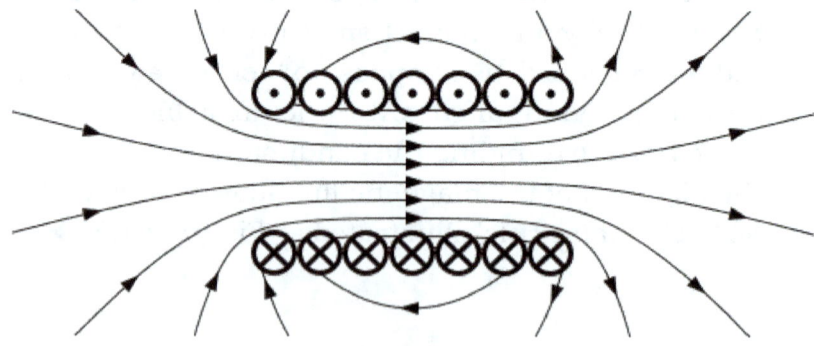

**Figure 2.1-8
Coil Turns and Magnetic Field Strength**

Electric Power Generation

Figure 2.1-8 (◄) is a cross-sectional view of a coil of wire. Current flows into the lower turns indicated by Xs and out of the upper turns indicated by dots. Using the right-hand rule, the direction of flux lines that surround each turn of the coil align to reinforce the other coil turns, adding field strength to the overall coil. Therefore, the field strength of a coil is the product of the number of its turns (N) and the current (A) flowing through it. The field strength of a coil may be expressed as follows using (Φ) to indicate magnetic field strength:

$$\Phi = NA$$

Coils take on the properties of a permanent magnet with North and South poles. Magnetic lines of flux lines align out of the North pole and into the South pole. Coils also provide more induced voltage in the stationary windings of the machine than single wire loops as indicated in Figure 2.1-4 because the voltage induced is also multiplied by the number of coil turns. The stationary coils surrounding the rotor of a generator are commonly referred to as stator windings or stators.

Electric Power System Fundamentals

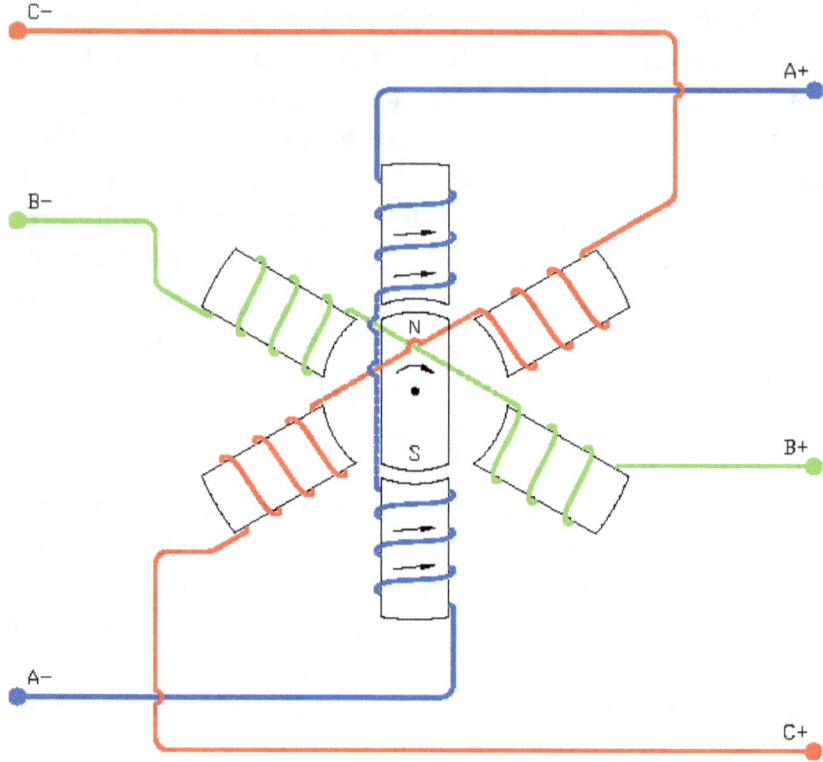

Figure 2.1-9
Generator Winding Details

Figure 2.1-9 (▲) illustrates the stator winding scheme of a three-phase commercial generator. The rotor is also comprised of a coil although the details are not shown. The stators are wound in pairs designated (A), (B), and (C) and wired together through the center of the machine to induct maximum current from the North and South rotor poles simultaneously. For example, with the rotor "frozen-in-time" as indicated, maximum potential is produced between the lower connection, A (–) and the

Electric Power Generation

upper connection A (+). If either half of the (A) winding pair were eliminated the induced voltage would be halved.

When the rotor is rotated at a constant speed, the blue, red, and green windings generate sinusoidal wave forms as illustrated in Figure 2.1-5.

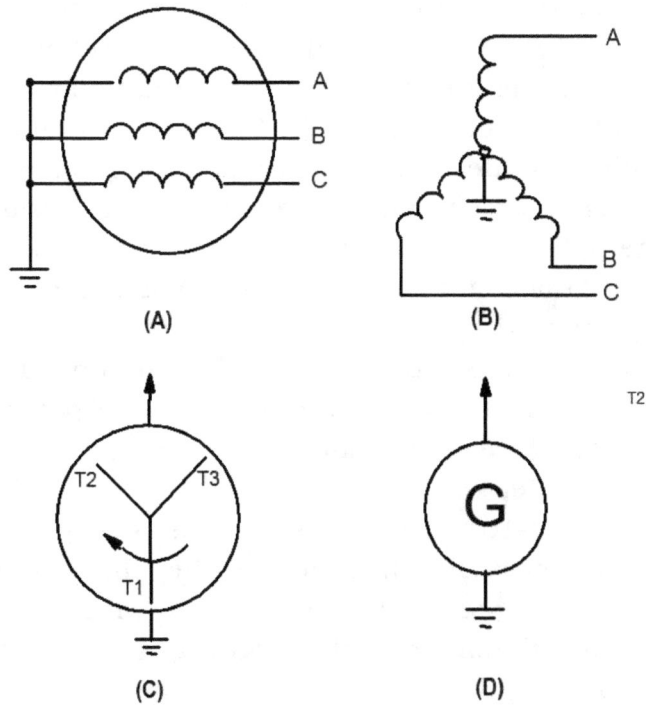

Figure 2.1-10
Generator Schematic Representations

Generators may be schematically represented in several ways. Four examples are illustrated in Figure 2.1-10 (▲).

The (A) diagram is a logical generator representation

because the stator winding ends are often physically located on the left-hand and right-hand sides of the generator enclosure, allowing the user to select the ground (neutral) connected side and line (output) connected side based upon convenience. If the rotor is turning clockwise, the (A) phase will precede the (B) phase followed by the (C) phase. If the opposite side of the machine is grounded and the direction of the machine rotor is the same, the phase rotation will be (C) followed by (B) followed by (A). The phase rotation can be changed as desired by external wiring connections regardless of which side of the generator is designated the line side. To avoid extreme damage to equipment, phase rotation must be matched to equipment that is already powered.

Synchronism equipment, described in Appendix B, is designed to assure phase rotation and other parameters are correct before connection is made to equipment that is already powered.

Figure 2.1-10 (C) and (D) (◄) use a single line to represent generator connections rather than three and are therefore designated as one-line drawings. One line illustrations simplify complex drawings as discussed in Section 3.2.

Regulating current flow through the spinning rotor of a generator regulates the strength of the magnetic field surrounding the rotor as well as the emf induced into the machine's stators. Rotor current is commonly referred to as rotor excitation current or "generator excitation current."

Electric Power Generation

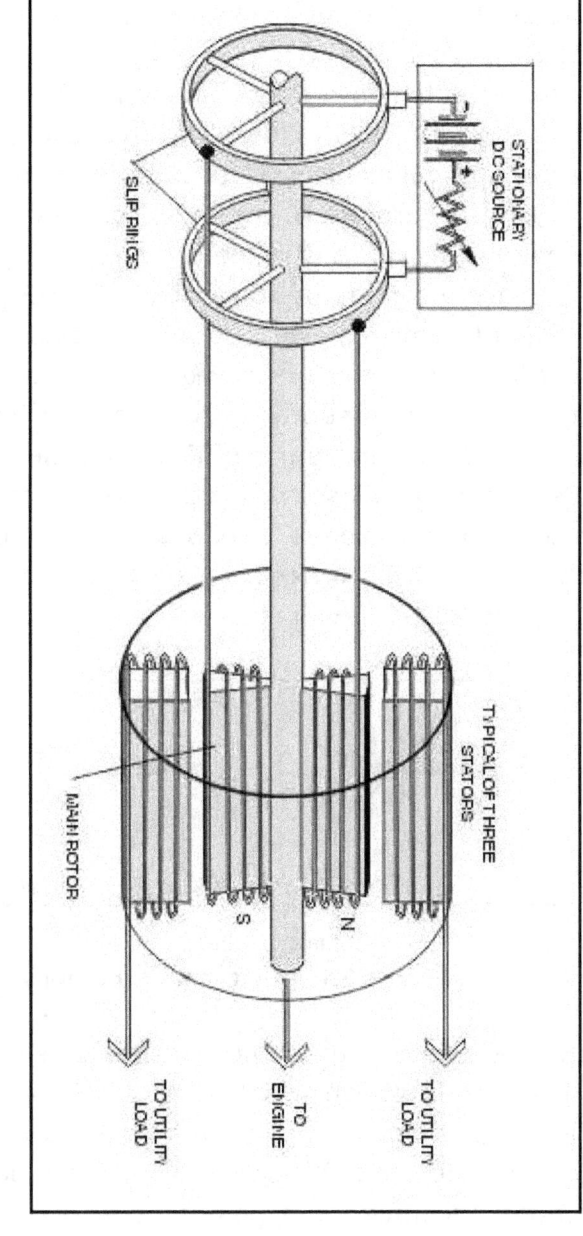

**Figure 2.1-11
Brush Type Generator Excitation Scheme**

29

Electric Power System Fundamentals

A so-called "brush" type exciter scheme is illustrated in Figure 2.1-11 (◄). Brush type exciters provide excitation current through a set of slip rings in contact with two brushes connected to a stationary current source. This scheme assures current flows through the rotor in the same direction regardless of its angular position. As in the action of a permanent magnet, the rotor North and South poles do not change polarity regardless of rotor angle. A variable resistor in series with the battery in Figure 2.1-11 provides excitation current control.

Brushless type exciters perform the same function as brush types but use magnetic coupling between the stationary and rotating parts of the machine rather than brushes. Brushless exciters also use semi-conductor components called "diodes" to convert alternating current to direct current where necessary on the stationary and rotating parts of the generator.

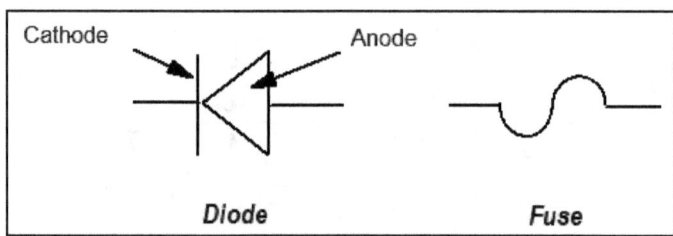

Figure 2.1-12
Diode Fuse Schematic Representations

Diodes are represented schematically as indicated in the left-hand illustration in Figure 2.1-12 (▲). Current flows through diodes only if the anode is at a greater voltage than the cathode. Current flow is blocked when the cathode becomes more positive than the anode. The

Electric Power Generation

point on the anode indicates the direction of conventional current flow through diodes, i.e. (+) to (–). Fuses are wired in series with diodes to protect other connected circuitry should diodes fail. Fuses are represented schematically as indicated in the right-hand illustration in Figure 2.1-12 (◄). A short length of fusible wire within a fuse opens when heated by current flow. The value of current sufficient to "burn" and open the fusible wire is a calibrated value. Fuses are selected for the value of current at which they open, preventing damage to connected circuitry. Diodes fail in one of two ways; either open or shorted. Should they fail open, current flow is prevented. Should they fail shorted, its associated fuse would open because of excess current flow.

A so-called brushless exciter scheme is illustrated in Figure 2.1-13 (►). Brushless type exciters provide rotor excitation without brushes. A permanent magnet of high permeability, or magnetic retention, is fixed to the end of the generator shaft. An alternating current is thereby induced into stationary windings adjacent to the rotating permanent magnet. The AC current generated is allowed to flow in only one direction through diodes on the stationary part of the machine, converting the AC current into DC current. The resulting DC current is applied through a variable resistor to a set of stationary coils called exciter field windings. A set of three windings, each spaced at 120 mechanical degree intervals on the rotor of the machine are induced with a current whose amplitude is regulated by the variable resistor. The output of the three exciter rotor windings is applied to a set of diodes on the rotor to produce DC current through the main rotor. The main rotor excitation current is regulated by the variable resistor.

Electric Power System Fundamentals

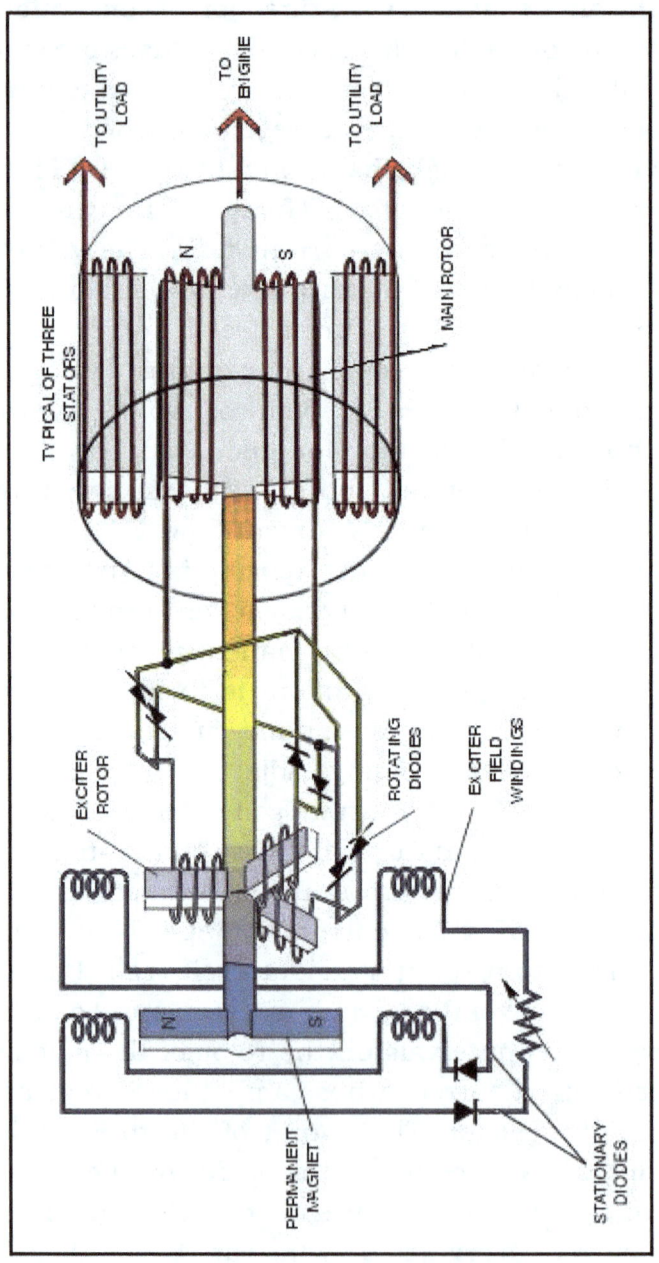

**Figure 2.1-13
Brushless Type Excitation Scheme**

Electric Power Generation

Brushes in contact with the rotating parts of a generator wear and require replacement. Brushes made from various materials also release a fine powder as they wear that must be isolated from other generator components. Brushless exciters do not share these problems but are not as responsive to extremely rapid excitation current changes as brush-type exciters.

The schematic in Figure 2.1-14 (▶) represents the moving components on a typical brushless exciter. The three-phase exciter coils are evident at the left-hand side of the drawing. The AC output voltage from each of the three excitation coils is applied through fuses to pairs of diodes. Because of the action of the diodes, only the positive half cycles of the three-phase exciter coils will appear on the positive side of the main rotor coil and the negative half cycles on the negative side. Should a diode fail or a fuse open, the sudden drop in excitation current would be detected by sensing circuitry in the excitation current regulator resulting in a "diode fail" signal to system operators.

The Voltage Threshold Detector connected to the ends of the main rotor coil provides an output signal should the voltage measured across the ends of the rotor fall below a pre-determined minimum threshold level. Depending upon the manufacturer, the threshold detector may interrupt a radio signal transmitted from the rotor or a Light Emitting Diode (LED) that is mounted on the rotor. Loss of these constantly monitored signals is interpreted as a "rotor ground fault." Rotor ground faults typically initiate a generator shut-down sequence to prevent permanent damage caused by excessive current through the main rotor windings.

Electric Power System Fundamentals

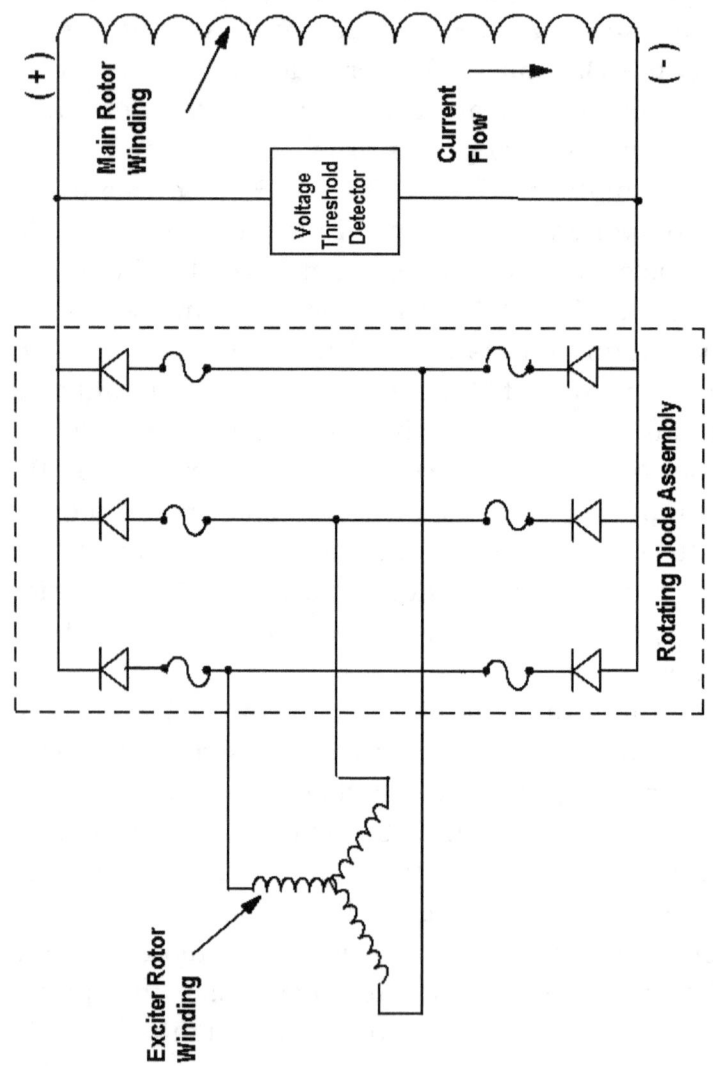

Figure 2.1-14
Brushless Generator Rotor Wiring

Electric Power Generation

Large power generators require lubrication, vibration detection, ventilation and cooling support systems. The major functions of these systems are discussed below.

<u>Lubrication systems</u> – Generator rotor components typically weigh tons and require bearings that absorb substantial quantities of heat caused by friction. For optimum protection, bearing lubricants are micro filtered and temperature and pressure regulated. Some machines require very high and concentrated lubrication pressures below the support bearings during start up to reduce the torque necessary to rotate the shaft after it has been stopped. Replacing generator bearings is very expensive in down-time revenue loss and machine repair cost; therefore, various redundant lubrication schemes are used to prevent bearing damage should pressure pumps, cooling systems, or filtration systems fail. Instrumentation is installed to monitor bearing and lubrication oil temperatures, lubrication pump pressures, and filtration quality. Operator alarms are initiated when critical limits are exceeded. In some cases, automatic shut-down sequences are initiated to avoid bearing damage.

<u>Vibration detection and monitoring systems</u> – The rotating components of a generator can become unbalanced, creating destructive levels of vibration. The most frequent cause of vibration is the shifting of coil turns on the machine rotor caused by the centrifugal force of rotation. Electrical sensing devices called "proximators," installed at 90 degree angles within the stationary bearing sleeves, measure the displacement between the bearing sleeve and the rotating shaft. A plot of the displacement with respect to time will produce

a straight line if the displacement is the same at every rotation angle. If the displacement is not the same at every angular shaft position a sinusoid is produced whose amplitude, calibrated in millimeters or in thousandths of an inch, indicates the level of vibration. Unless the vibration level is severe, counter weights can be installed on the rotor to counteract the motion. The direction of the motion is determined from the phase angular relationships between pairs of proximators installed at 90 degree angles. The size of counter weights installed to nullify the vibration relates to the amplitude of the sine wave. Vibration levels are displayed for monitoring and can initiate automatic shut-down of the machine if they exceed safe limits.

Cooling and ventilation systems – Heat is produced within the enclosures of generators from several sources. The principal source of heat is from electrical resistance within the stator windings of the machine that increase as the generator's electrical load increases. In air-cooled units, fans on each end of the generator rotor shaft take their suction from filtered air surrounding the generator enclosure and force it through passages around the rotor and exciter components. Temperature sensors imbedded in the stator windings and at the inlet and outlet air passages provide temperature monitoring and alarm signals. To avoid permanent damage, operators must reduce the generator's output load when stator temperatures approach their operational limits. Mechanically refrigerated air, water vaporization, and liquid cooling systems are optional depending upon generator size, electrical loading requirements, and

Electric Power Generation

other factors such as ambient temperature and relative humidity.

The machine illustrated in Figure 2.1-15 (▶) is typical of air-cooled generators in the 50 megawatt class. The major components are listed below:

1. Stator Windings – Stationary wire coils into which voltage is induced by the moving magnetic field surrounding the rotor.

2. Stator Core – Thin steel laminations are stacked together to form a core which supports the stator windings.

3. Rotor – A 12-ton solid forging of nickel-chromium-molybdenum alloy steel that supports solid copper bar rotor windings and rotating exciter components.

4. Rotor End-caps – The end-caps are non-magnetic steel covers that protect the end portions of the rotor windings.

5. Shaft-Mounted Fans – Three fans fixed to the rotor force cooling air through the exciter and stator windings and out through the top of the generator enclosure.

6. Pressure Oil Seals – Twin lube oil seals are mounted at the inner and outer edge of each bearing cavity to retain oil within the machine. Air pressure from the shaft fans is also inserted between the seals to assist in containing the bearing lubrication oil.

Electric Power System Fundamentals

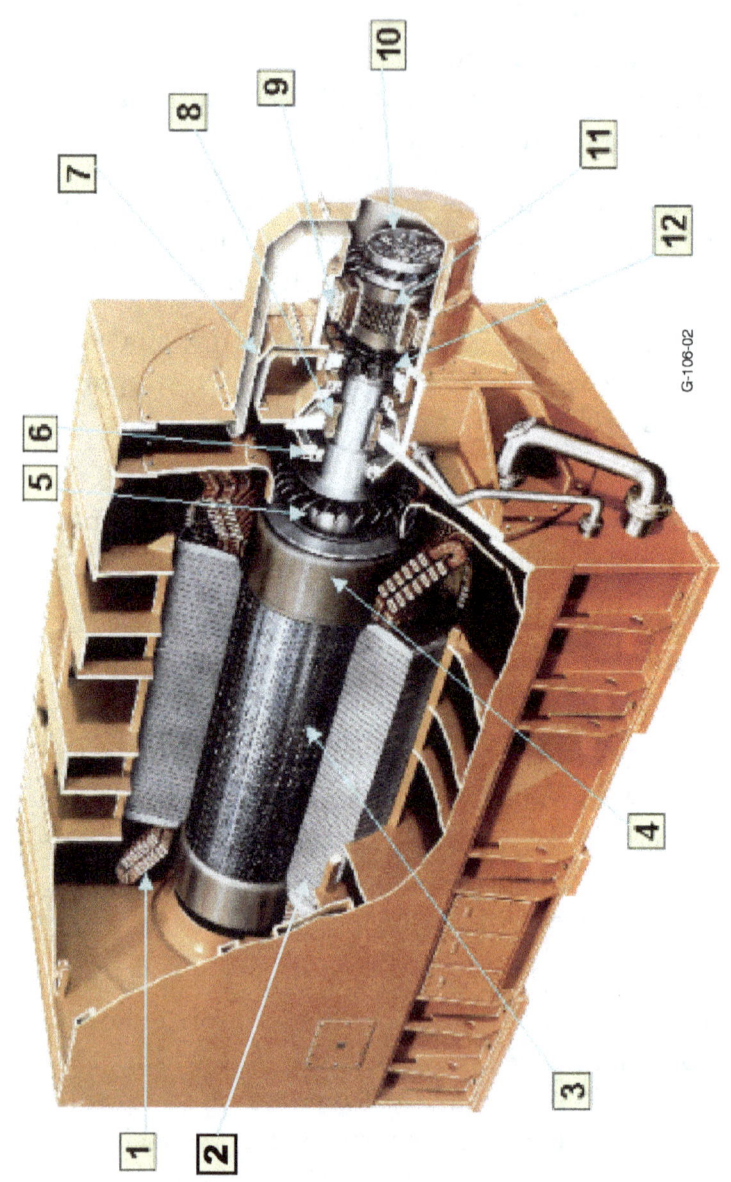

**Figure 2.1-15
50 MW Class Air Cooled Generator**

Electric Power Generation

7. Exciter Cooling Air Duct – One of the three shaft-driven ventilation fans pulls cooling air through this duct for cooling the exciter components.

8. Rotor Bearings – White metal-lined, cylindrical bearings support the rotor shaft at each end. These bearings require a continuous flow of filtered, pressure and temperature regulated lubrication oil when the rotor is turning.

9. Exciter Stator – Stationary coils through which DC current is regulated to produce a magnetic field. Rotating components of the brushless exciter convert the strength of this field into rotor excitation current.

10. Rotating Diodes – These diodes convert AC exciter current into DC current for application to the main rotor.

11. Exciter Rotor – Contains the rotating components of the brushless exciter.

12. Permanent Magnet Generator (PMG) – Generates AC current that is converted to direct current and then regulated and applied to the exciter stator windings.

**Figure 2.1-16
A Commercial Generator Rotor**

**Figure 2.1-17
Diode Wheel Assembly**

Electric Power Generation

A cast and machined rotor ready for winding and installation is illustrated in Figure 2.1-16 (◄). Insulated copper bars are wound in a continuous coil through the slots on each side of the rotor. The two ends of the winding are brought out through the bore in the shaft and become the positive and negative end points to which the exciter rotating diode assembly is connected. The bolt circle on the end of the shaft provides mounting for the diodes and other rotating exciter components.

The assembly illustrated in Figure 2.1-17 (◄) provides mounting for the rotating diodes and fuses schematically illustrated in Figure 2.1-14. The metal tab extending into the center of the assembly is one of the two connection points for the main rotor winding; the other is not illustrated. The wing-shaped extensions around the outer ring of the assembly are ventilation fan blades that accelerate the flow of cooling air when the rotor is turning.

Electric Power System Fundamentals

Section 2.1 Questions:

1. How is power plant efficiency expressed?
2. Name four important facts that were learned about electro-magnetism from Faraday's early experiments.
3. How is a mechanical pump moving liquid through a pipe analogous to Faraday's experiment with a magnet and an electrical conductor?
4. How does direct current differ from alternating current?
5. Why do coils of wire carrying an electric current produce larger magnetic fields that single straight wires?
6. Why do stator windings near the North and South poles of a generator rotor produce more voltage than a single stator winding adjacent to a single pole.
7. What are the advantages and disadvantages of brushless generator exciters compared to brush-type exciters?
8. Why are rotating diodes necessary in a brushless generator exciter?
9. What is the function of a prime mover in an electric generating plant?
10. At what speed should a single 2-pole (N-S) rotor turn to produce 50 Hz electric power?

11. What actions should an operator take if the stator windings of a generator approach their upper temperature operating limits?

12. What is meant by the abbreviation "emf?"

Notes

Electric Power Generation

2.2 GENERATOR CONTROLS

An alternating current generator has essentially two external controlling mechanisms:

1. regulating the speed or power applied to the generator rotor from a prime mover, and

2. regulating the rotor excitation current.

Each of these controls affects the generated power output in distinctly different ways and will be discussed separately.

2.2.1 REGULATING THE PRIME MOVER THROTTLE

As we have seen, the speed of rotation of a generator rotor determines alternating current frequency and from Faraday's demonstration we have also seen that mechanical work is needed to produce an electric current; therefore, when electrical load is increased, more shaft rotational power is needed to maintain the same shaft speed. To appreciate how power plant operators regulate power requires a definition of electrical power and mechanical power and how they are related.

2.2.1.1 Electric Power Definition

German Physicist, Georg Ohm, discovered that different amounts of current flowed between battery terminals when they were connected through different lengths of wire. He attributed the difference in current flow to a change of what he termed *resistance* to current flow through the different lengths of wire. From his studies, Ohm related voltage, current, and resistance, into what has become known as Ohms Law expressed as follows:

$$V = AR \quad \text{or} \quad A = \frac{V}{R} \quad \text{or} \quad R = \frac{V}{A}$$

where,

V = Voltage or electro motive force in volts

A = Current in amperes, and

R = Resistance, the unit of which became known as the ohm

The importance of this law cannot be overstated in any study of electricity. It explains the relationship between voltage, current, and resistance in all electrical circuitry, stating that the amount of current that flows through an electric circuit is directly proportional to the applied voltage and inversely proportional to the circuit's resistance.

Independent discoveries by Georg Ohm and James P. Joule and later proved equivalent by James C. Maxwell allow writing the following expressions that define electrical power.

Ohm's Law:

$$V = AR \quad \text{or} \quad A = \frac{V}{R} \quad \text{or} \quad R = \frac{V}{A}$$

Electric Power Generation

Joule's Law:

$P = A^2R,$

where,

P = power

Combining Ohm's and Joule's Laws:

$V = AR$, and $R = \dfrac{V}{A}$ then,

$\dfrac{A^2V}{A} = VA$ and, $A^2R = VA$

Power = $A^2R = VA$

where,

P = electric power in watts

V = voltage in volts

R = electrical resistance in ohms

$I = A$ = current in amperes

<u>Note:</u> Voltage is designated throughout this text as V or E; current is designated as I or A.

The expression, $P = VA$, has three algebraic forms:

$$P = VA \quad \text{or} \quad EI = \dfrac{E^2}{R} = I^2R$$

47

2.2.1.2 Mechanical and Electrical Power

Mechanical power can be expressed as an amount of work expended over an interval of time and work as an amount of force in pounds times distance:

$$\text{Power} = \frac{\text{work}}{\text{time}} = \frac{\text{force} \times \text{distance}}{\text{time}}$$

Units of power in English units are expressed in foot-pounds per unit time. A familiar unit of power is the horsepower defined as 33,000 foot-pounds of work per minute.

As we have seen, the motion of a coil through a magnetic field requires an expenditure of energy to produce a resultant electromotive force (emf) or voltage, causing current to flow in an electric circuit in proportion to the resistance it encounters as described by Ohm's law. Early experimenters also knew that current flowing through a conductor produced heat. English scientist James P. Joule further realized that if mechanical power could be related to heat and heat to electrical power, then mechanical power could be related to electrical power.

Electric Power Generation

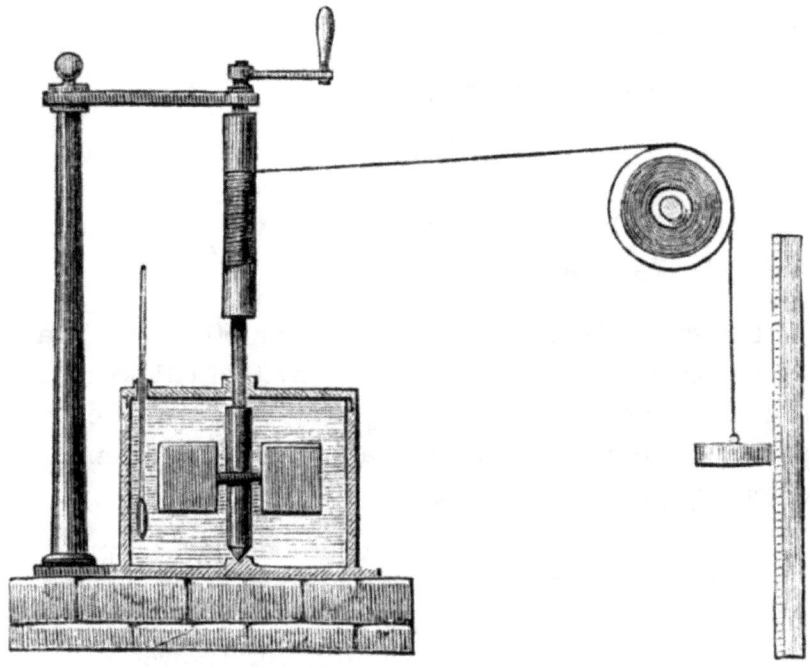

Figure 2.2.1.2-1
Joule's Work - Heat Demonstration

Using the primitive apparatus illustrated in Figure 2.2.1.2-1 (▲), Joule demonstrated that a relationship existed between mechanical work and heat.

Turning the crank did an amount of mechanical work equal to the distance the weight was raised in a unit of time. The rotating paddle wheel heated the water in the tank, causing the temperature to increase as measured by the thermometer submerged in the water. Joule then arbitrarily established the *joule* as the unit relating the work required to move units in grams per meter per second defining it such that 1 *joule* per second equaled

1 watt. To make this sensible in English units, 1.356 joules equals 1 foot-pound, or:

$$1 \text{ ft-lb/sec.} = 1.356 \text{ joules} = 1.356 \text{ watt/sec.};$$

therefore:

$$1 \text{ HP} = 550 \text{ ft.-lbs/sec/hp} \times 1.356 \text{ watts/ft-lb/sec.}$$
$$= 746 \text{ watts/hp}$$

For example, a load of 50 MW or 50 million watts would require $50 \times 10^6/746$ HP or 67,024 HP be applied to the rotor of a generator to produce 50 MW of electric power assuming 100% generator efficiency. Generator efficiency may be expressed as follows:

$$\text{Generator Efficiency} = \frac{\text{Mechanical Power Input}}{\text{Electrical Power Output}}$$

$$= \frac{\text{Driving Horsepower}}{\text{Watts Generated}}$$

$$= \frac{\text{HP}}{\text{W}}$$

Generator efficiencies in the order of 98% are not uncommon; therefore, except in precise calculations, generator efficiency is neglected and not considered significant.

2.2.2 REGULATING ROTOR EXCITATION CURRENT

Regulating current flow through the spinning rotor of a generator regulates the strength of the magnetic field surrounding the rotor as well as the emf or voltage that is induced into the machine's stators. When

Electric Power Generation

external loads are connected, the generator's reaction to excitation current is affected by the characteristics of the loads.

In alternating current circuits, current flow is impeded by three components: inductive reactance, capacitive reactance, and resistance. When these elements are combined, they comprise what has become called impedance. Impedance resists current flow in alternating current circuits just as resistance alone does in direct current circuits. Inductive reactance, capacitive reactance, resistance, and impedance are defined in the following paragraphs.

2.2.2.1 Inductive Reactance

It can be demonstrated that a finite time is required for magnetic fields to reach their maximum strength after voltage is first applied. Likewise, when the applied voltage is removed, a finite time is required for magnetic fields to collapse. Because of this phenomenon, current flow through an inductor in AC circuits is delayed 90 degrees following the applied voltage. This behavior is called *inductive reactance* and is denoted as X_L. The magnitude of reactance through a coil will be greater than through a single conductor because the coil has a larger field that must build and collapse. Coils, therefore, offer greater *inductive reactance* than single conductors. The property of inductance is designated "L" whose strength in designated in "Henrys." The value of inductance increases with the rate of change of the applied voltage because the magnetic field surrounding an inductor must build and collapse more rapidly. The inductive reactance (X_L) in a circuit is therefore proportional to the frequency of the applied voltage and to the value of the circuit's inductance and may be expressed as:

$$X_L = 2\pi f L$$

where,

X_L = Inductive Reactance in Ohms,

f = Frequency of the applied voltage in cycles per second, and

L = Inductance in Henrys

2.2.2.2 Capacitive Reactance

In the earliest studies of electricity, scientists were concerned with what is known as "static (or stored) electricity." The device in which static electricity is stored is called a condenser or a capacitor. A simple capacitor consists of two metallic plates separated by a dielectric insulating material. In ordinary engineering practice, a capacitor takes the form of sheets of metal separated by insulating material rolled into a cylindrical shape or a flatter shape for very small capacitors. When voltage is applied across the plates of a capacitor the plate attached to the positive voltage terminal immediately acquires a positive charge and the plate attached to the negative terminal immediately acquires a negative charge. The instant current flow created appears as a short, dropping the voltage applied across the plates until they have time to recharge to equal the applied voltage. Current flow precedes the *voltage* across capacitors by 90 degrees. This behavior is called *capacitive reactance* and is denoted as

X_c. Capacitive reactance decreases as the applied voltage frequency increases because a large capacitor requires a longer charging time than a smaller one. The property of capacitance is denoted as "C" and the unit of capacitance is denoted in "Farads." The capacitive reactance in a circuit (X_c) is therefore inversely proportional to the frequency of the applied voltage and the capacitance of the circuit (C) and may be expressed as follows:

$$X_c = \frac{1}{2\pi fC}$$

where,

X_c = Capacitive Reactance in Ohms

f = Frequency of the applied voltage in cycles per second, and

C = Capacitance in Farads

2.2.2.3 Resistance

Resistance is that property of electrical loads that does not delay current after voltage is applied as do inductors or cause the applied voltage to follow current flow as is the case in capacitive circuits. Ohms law describes the physical nature of resistance as R = V/A and its characteristics do not change in alternating current circuits.

2.2.2.4 Impedance

Impedance is the name given to inductive, capacitive, and resistive elements when they are combined in electrical circuits. In alternating current circuits, *impedance* is

Electric Power System Fundamentals

equivalent to *resistance* in direct current circuits. The property of impedance is denoted as "Z" and the unit of impedance is the ohm. Impedance may be expressed consistent with Ohm's Law as follows:

$$Z = \frac{V}{A} \quad \text{or} \quad A = \frac{V}{Z} \quad \text{or} \quad V = ZA$$

Although impedance resists current flow in alternating current circuits, resistance is the only component that absorbs power.

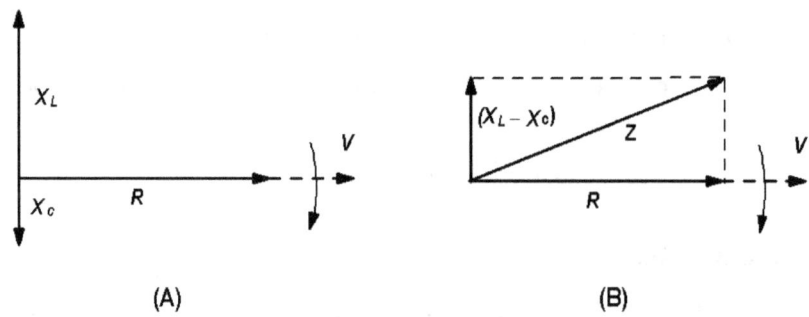

Figure 2.2.2.4
Impedance Diagrams

Because X_L and X_C, lag and lead the voltage applied to them respectively and R occurs at the same time voltage is applied, Z can be illustrated diagrammatically as illustrated in Figure 2.2.2.4 (▲). The quantity of resistance (R) is denoted by a horizontal vector that occurs at the same time and is in-phase with the applied voltage V. In Figure 2.2.2.4 (A), the vector quantity X_L lags behind the clockwise rotating vector V and X_C leads the voltage vector. Because the inductive and capacitive reactance vectors diametrically oppose one another, they may be algebraically added resulting in the (B) example in which case the overall circuit characteristic is inductive because X_L is greater than X_C. The impedance of the circuit (Z) then becomes the vector sum of R and X_L as indicated. Current is maximum in an alternating current circuits when $X_L = X_C$ resulting in Z becoming equal to R. In power systems this means maximum power is transferred to the load in the absence of a reactive component.

Because a generator is in series with the loads it supplies power, and its excitation current induces inductive reactance into its stator windings, a generator contributes inductive reactance into the circuit it is supplying in proportion to its excitation current. This implies that if a generator is supplying an inductive load, it should be operated with the lowest possible rotor excitation current and if it is supplying a capacitive load it should be producing an amount of inductive reactance that minimizes the capacitive reactance of the load. There is a *minimum* excitation current limit below which the rotor would contribute insufficient electro-motive force to maintain an output, in which case, the rotor would accelerate with the loss of load. Such a condition is called "pole slip." The *maximum* allowable excitation current is limited by overheating of generator rotor components. These safe limit boundaries are summarized in the generator capabilities curve illustrated in Figure 2.2.4 (▶).

Electric Power System Fundamentals

2.2.3 CONTROL EFFECTS ON GENERATED POWER

As we have seen, power consumed in direct current circuits can be expressed as P = VA. In AC circuits, however, current (A) is determined by the impedance (Z) of a circuit which is comprised of inductive and capacitive reactance components as illustrated in Figure 2.2.2.4 (◄). This means that VA is only the apparent power transferred from an AC generator to its load and does not consider the effects of the inductive and capacitive load components. Finding the real power transferred from a generator to its load is a complex problem for which power companies use a short cut that determines real power instantly. The methodology is described graphically as follows:

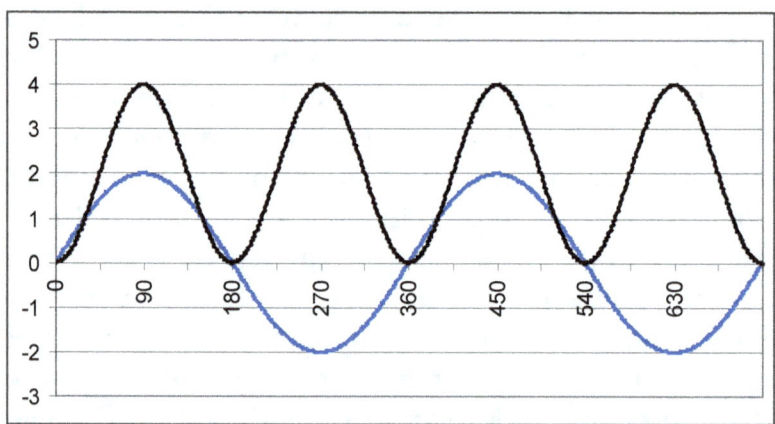

**Figure 2.2.3-1
Voltage and Current In-Phase**

The blue waveform in Figure 2.2.3-1 (▲) combines voltage and current when they occur at the same time. The black curve illustrates real power, the product of each instantaneous value of voltage and each instantaneous value of current. In this diagram the peak value of power reaches a maximum of 4 units on the graph. The average power is, therefore, the maximum value (4) divided by 2 or 2 units.

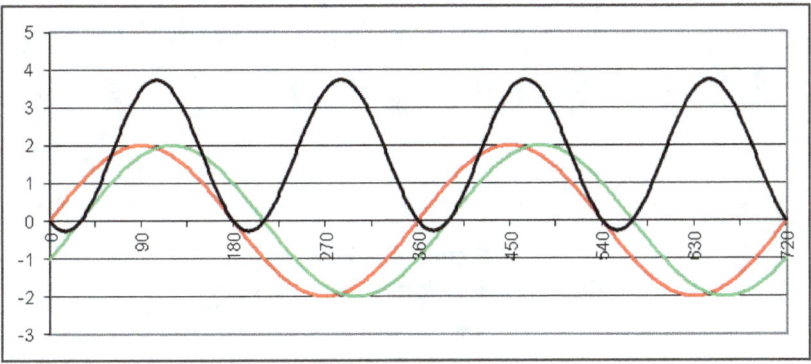

**Figure 2.2.3-2
Current Lags Voltage 30 Degrees**

In Figure 2.2.3-2 (▲), voltage is illustrated in red and current in green. Because current follows the applied voltage caused by inductive reactance, real power drops below its maximum value. The greater the delay in current caused by increased inductive reactance, the greater the loss of power. In Figure 2.2.3-2, the delay is 30 degrees. Power drops below the zero level until both current and voltage are positive or negative. When that occurs their product is again positive. (i.e., The product of two positives or two negatives is positive).

Electric Power System Fundamentals

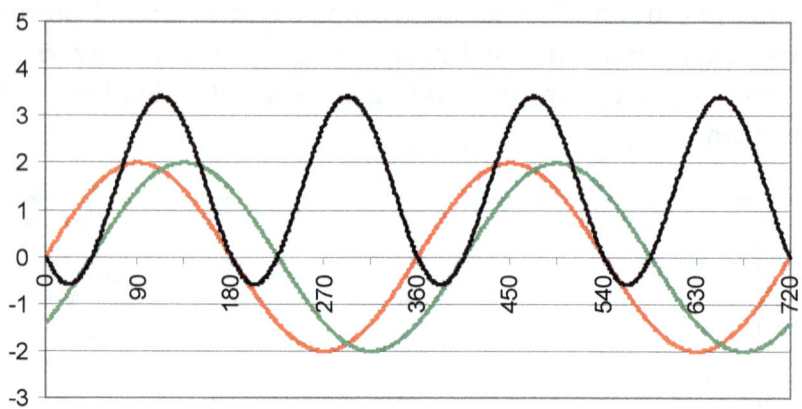

**Figure 2.2.3-3
Current Lags Voltage 45 Degrees**

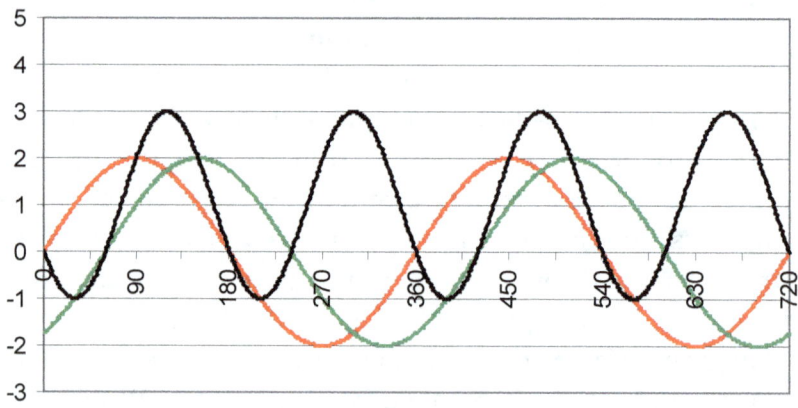

**Figure 2.2.3-4
Current Lags Voltage 60 Degrees**

Electric Power Generation

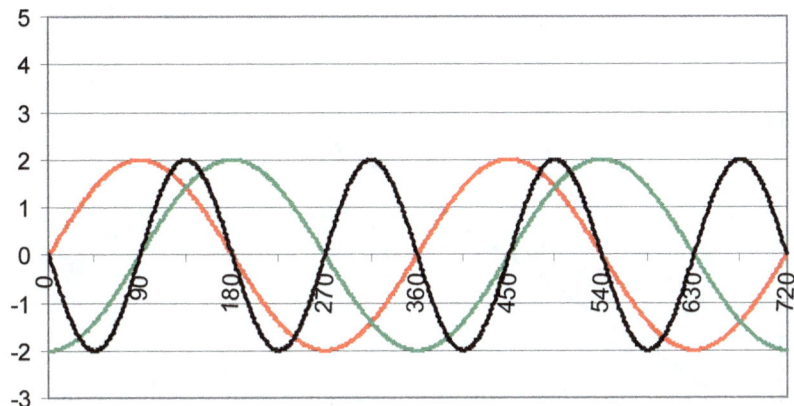

**Figure 2.2.3-5
Current Lags Voltage 90 Degrees**

Power continues to drop as the delay is increased to 45 and 60 degrees and reaches zero when the delay is 90 degrees as illustrated above.

If the load became capacitive rather than inductive, voltage would occur after current with the same effect on average power. The capacitive case has not been detailed because most industrial and residential loads are inductive.

Electric Power System Fundamentals

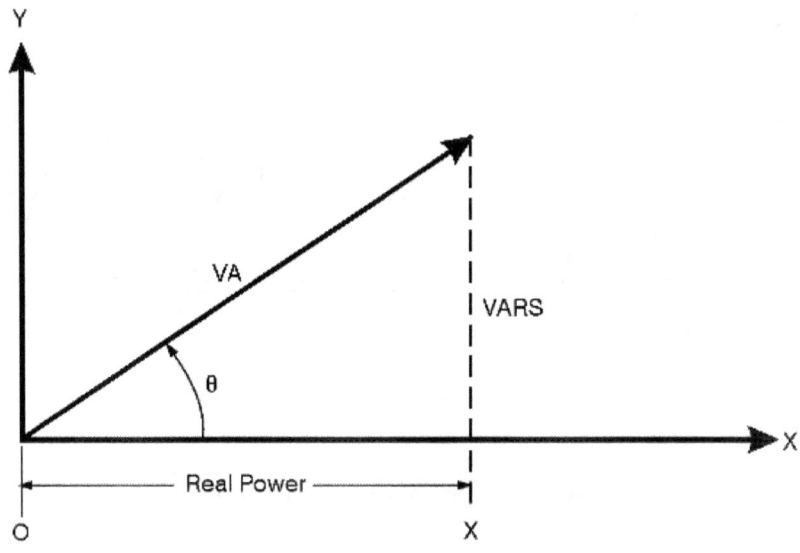

Figure 2.2.3-6
The Power Triangle

Average power could be found by the graphical method, however, a much faster and more practical method has been developed using the mathematical model illustrated in Figure 2.2.3-6 (▲). Given that current and voltage values are immediately available from current and potential transformer instrumentation as well as voltage and current angular displacement in real time, the product of measured voltage (V) and current (A) can be illustrated as a vector placed at an angle (θ) equal to the measured angle between current and voltage. To complete the model, a vertical line is constructed between the terminus of the VA vector and the x-axis. The quantity OX is then defined as *real power*.

As the current-voltage angle increases due to

reactance, real power will decrease and, as the current voltage angle approaches zero degrees, real power approaches the value of VA. Since the cosine of the voltage current angle is real power divided by VA, real power is then VA (Cos θ) as follows:

$$\cos \theta = \frac{\text{Real Power}}{\text{VA}} \quad \text{or} \quad \text{VA}(\cos \theta) = \text{Real Power}$$

The cosine of the voltage-current angle θ is defined as the *Power Factor* and the imaginary side of the triangle opposite θ is defined as volt-amperes-reactive or VARS. VARS are also instantly calculated from measured data as VA(Sin θ).

Thousands of volt amperes, real power, and reactive volt amperes are expressed as kVA, kW, and KVARS. Millions are expressed as MVA, MW, and MVARS.

When connected loads are capacitive rather than inductive, current *leads* the applied voltage causing the angle θ to fall below the horizontal axis (x) in which case power will remain positive but VARS will become negative. VARS are positive for inductive loads and negative for capacitive loads. Although VARS are an imaginary quantity, when VARS are positive the generating system, in the parlance of power plant operators, is said to be *making* VARS and when VARS are negative, the generating system is said to be *absorbing* VARS.

2.2.4 GENERATOR SAFE OPERATIONAL BOUNDARIES

Figure 2.2.4 (▶) represents a typical generator capabilities diagram. A power triangle is superimposed on top of curve to illustrate the generator's operating point (X) at the tip of the VA vector. Should the operating point move outside the grey area, generator damage could result. Although automatic protection circuitry is in place to prevent such an occurrence, operators should be constantly aware of the location of the operating point to avoid generator shut-down.

The right-hand boundary (A) of the curve defines the upper rotor excitation limit beyond which rotor excitation current could over-heat the rotor coil windings or other rotor exciter components. If load characteristics and the generator's excitation current cause the operating point to move outside the right-hand boundary of the capabilities curve, the machine is said to be *over-excited* and excitation current should be lowered to bring the operating point back within the safe limit boundary.

Real power is represented on the vertical axis, consequently, if real power exceeds the upper boundary (B) of the curve, the machine is producing more power than design limits allow. The real power boundary is limited by the current carrying capability of the machine's stator windings. As excessive current heats the windings their internal resistance is increased creating more heat by the I^2R rule.

Electric Power Generation

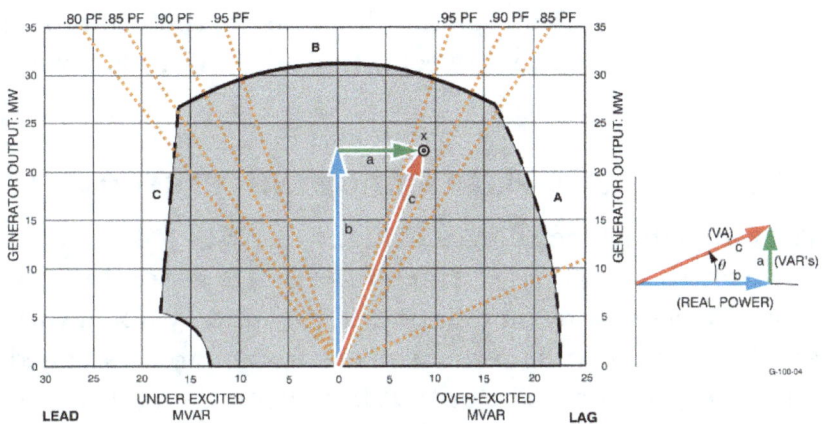

Figure 2.2.4
Typical Generator Capabilities Diagram

The left-hand boundary (C) defines the rotor under-excitation limit below which the rotor shaft horsepower could exceed the minimum magnetization current for power generation in which case the rotor speed would exceed the power frequency speed as if the prime mover had lost its load. This condition called "pole slip." If load characteristics and the generator's excitation current cause the operating point to move outside the left-hand boundary of the capabilities curve, the machine is said to be *under-excited* and excitation current should be increased to bring the operating point back within the safe limit boundary.

In summary, varying the generator rotor excitation current increases or decreases generator inductive reactance, thus adding-to or lowering the self-inductance of the generator. Increasing the generator excitation current will move the VA vector in the clockwise direction because the generator's inductance is in series

63

with the loads it is supplying. Decreasing the excitation current will move the VA vector in the counter-clockwise direction.

2.2.5 CONTROL IMPLEMENTATION AND OPERATION

The functional block diagram, Figure 2.2.5 (▶), illustrates how controls are implemented for a single generator in a typical power plant installation. The round symbols with an "X" in the center represent electronic circuits called *summing junctions*. The output from a summing junction (c) is the sum its inputs at (a) and (b) such that c = a + b. A negative (–) sign at an input indicates the polarity of that particular input will be inverted at (c). The (a) inputs above are each associated with a (–) sign; therefore, the (c) outputs are expressed as (c) = (b) – (a).

With the operator selected switch positions as shown, the upper switch is feeding the generator's measured *frequency* into its associated (a) summing junction input. The *set point* input at (b) will then determine the *frequency* of the generator as follows:

If the measured frequency at (a) is less than the set point at (b), (c) will be positive (c = b + a) signaling the prime mover throttle to increase speed. If the measured frequency at (a) is greater than the set point at (b), (c) will be negative (c = b – a) signaling the prime mover throttle to slow the prime mover speed. The action of the summing junction will drive the prime mover throttle to equal the input at (b); therefore, the set point position will determine the speed of the prime mover.

Electric Power Generation

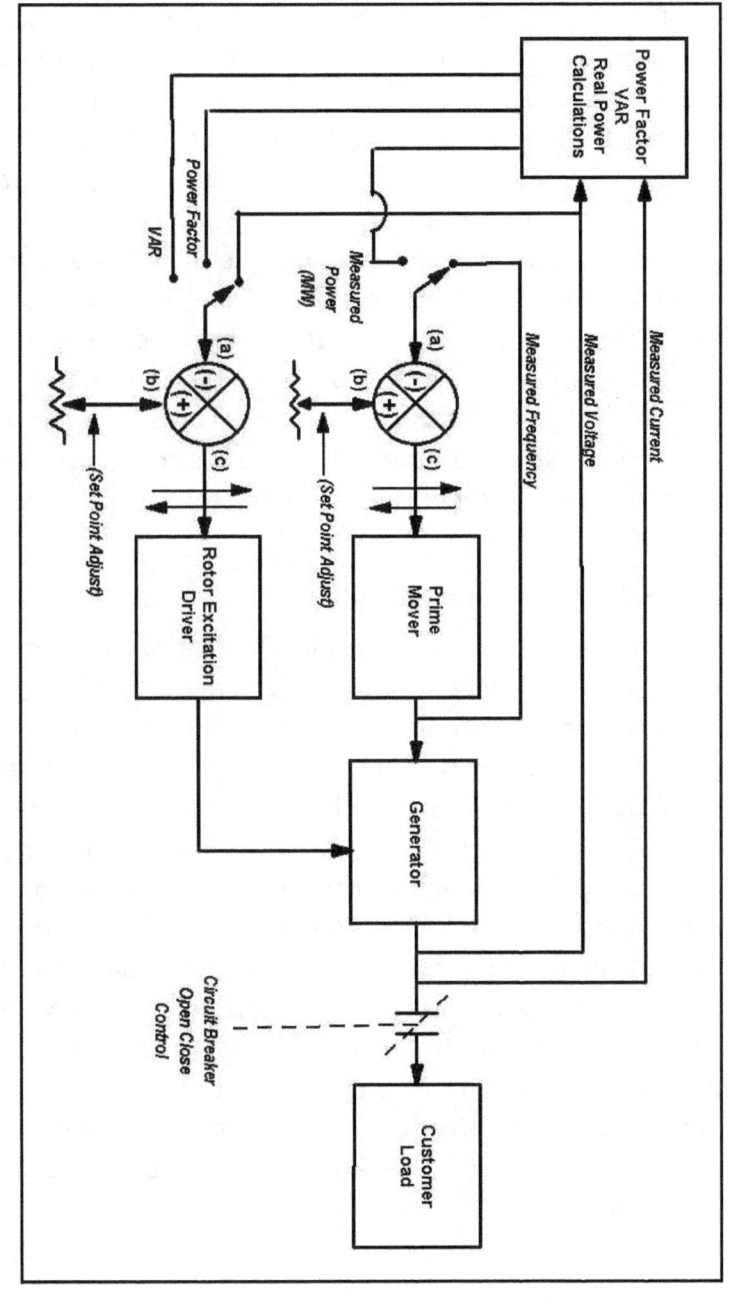

**Figure 2.2.5
Operator Control Implementation**

Electric Power System Fundamentals

The same action occurs at the lower summing junction with its selector switch set to apply the measured generator output voltage to the (a) summing junction input, driving the generator rotor excitation current up or down until the generator output voltage equals the voltage set point value.

Generator controls are set as indicated when the generator output circuit breaker is open to allow adjusting the voltage and frequency before connecting it to loads. If the generator is to be connected to lines that are already powered, a synchronizing procedure is required before closing its output circuit breaker. The synchronizing procedure is outlined in Appendix B.

2.2.5.1 Isochronous Control

If the generator is powering loads that are not connected in parallel with other generators it is said to be operating in isochronous mode, in which case it establishes the power frequency and maintains the desired voltage.

When operating in isochronous mode, load increases act as a brake on the generator shaft requiring more and more power from the prime mover to maintain the desired power frequency. Likewise, more rotor excitation current is required to maintain the desired voltage output. Eventually limits will be reached by the capability of the prime mover or the generator to produce more power. Safety overload devices will then alert operators to reduce load or in extreme cases open the output circuit breaker to prevent damage to the generator.

Electric Power Generation

2.2.5.2 Parallel Control

The parallel connection of many generators sharing interconnected loads has created complex networks that have become known as power grids. An infinite grid is generally considered one in which the power contributed by a single generator is not greater than $1/20^{th}$ the total power supplied to the network. This implies that each generator contributing to an infinite grid represents no more than $1/20^{th}$ or 5% of the parallel impedance of the network. Therefore, the self-impedance of a single generator is 95% or larger than that of the connected grid and changing its rotor excitation current does not significantly raise or lower its output voltage. Changes in excitation current increase or decrease the magnetization induced into the stator windings, resulting in changes in the self inductance of the machine with corresponding changes in delaying or advancing voltage with respect to current. As we have seen, power factor and VAR values are calculated from the trigonometric Sine and Cosine of the angular displacement of current and voltage.

The control diagram illustrated in Figure 2.2.5-1 includes a functional block in the upper left-hand corner in which power factor, VAR, and real power values are derived from measured values of voltage and current. The power factor or VAR calculated values are selectable inputs to the rotor excitation summing junction, allowing operators to set the desired VAR or power factor control points.

Figure 2.2.5-2 (▶) illustrates two versions of the power triangle. The diagram on the left is one in which the VAR level is maintained at a constant value as the load (VA) changes. In this case, the power factor angle (θ) also changes. The diagram on the right is one in which the power factor angle (θ) is fixed as VARs and the load (VA) change. As previously discussed, regardless of whether VARs are fixed or the power factor angle (θ) is fixed, real power is calculated as VA Cos (θ).

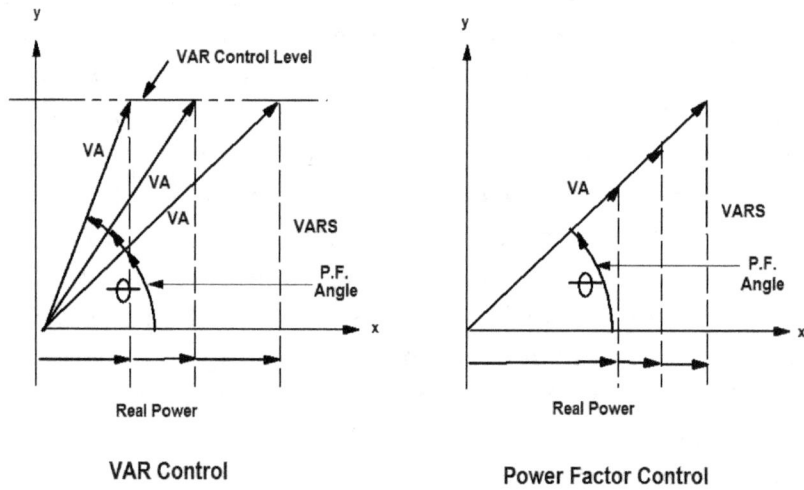

**Figure 2.2.5-2
VAR Power Factor Control Options**

Table 2.2.5 (▶) summarizes the results of regulating prime mover and rotor excitation current controls in isochronous and parallel modes of operation.

Electric Power Generation

CONTROL INPUTS	ISOCHRONOUS MODE	PARALLEL MODE Assumes an Infinite Grid
\multicolumn{3}{c}{GENERATOR REACTIONS to CONTROLS when in ISOCHRONOUS or PARALLEL MODES of OPERATION}		
Increase / Decrease Prime Mover Throttle Position	Increases or Decreases Generator Output Frequency	Increases or Decreases Generator Power Output
Increase / Decrease Gen. Excitation Current	Increases or Decreases Generator Output Voltage	Increases or Decreases Power Factor / VAR Values

Table 2.2.5
Generator Control Reactions

Electric Power System Fundamentals

Section 2.2 Questions:

1. What external controlling mechanisms are available for controlling a generator?
2. How did Joule establish a relationship between power and heat?
3. How is generator efficiency expressed?
4. What is the difference between impedance and reactance?
5. Does capacitive reactance lead or lag the applied voltage?
6. Does inductive reactance lead or lag the applied voltage?
7. Under what condition is impedance Z minimized?
8. When VARS are positive, is the associated load inductive or capacitive?
9. What action can an operator take to decrease VARS when the generator load is capacitive?
10. How is the operating point of a generator determined?
11. After a generator is connected to an infinite grid, how is its output power regulated?

Notes

Notes

3

ELECTRIC POWER TRANSMISSION AND DISTRIBUTION

Arguably the most important aspect of Tesla's concept of an alternating current system was its ability to transmit electric power over long distances with minimum loss. This is accomplished by using transformers to raise the level of generated voltages to high levels for transmission and then lower them for local distribution. As we will see, a tenfold increase in transmission voltage produces a one-hundred fold decrease in power losses through transmission lines.

High voltage transmission lines require expensive land right-of ways and safe clearance distances above ground, structures, people, and animals. High voltage transformers, measurement instrumentation, and maintenance servicing is also more expensive than for lower voltage systems.

The losses in lower voltage transmission lines of 40-to-60 miles in length are not appreciable compared

73

to the savings realized when transmission distances are greater than 40-to-60 miles and voltages above 100 kV provide realistic economic trade-offs.

For these reasons, power from generation plants is typically billed to transmission companies who in-turn bill distribution entities for the power they sell directly to end users. This three-tiered billing structure for power generation, transmission and distribution is more efficient because of the natural operational expenses incurred at each level. For example, it would not be sensible for a distribution company to be concerned with managing the heat rate of a power plant, nor would it be sensible for a power transmission company to be concerned with the maintenance, up-keep, and billing of thousands of customers served over many widely displaced distribution points.

The generally accepted division between transmission and distribution level voltages is 100 kV. Common transmission voltages above 100 kV are 115 kV and 225 kV although higher voltages in the range of 1,200 kV are in use. At extremely high voltages above 2,000 kV corona or arcing discharge losses can offset the economies gained by transmitting power at lower voltage levels.

Distribution voltage levels in common use are 69 kV, 39 kV and 10 – 15 kV although these levels may vary depending upon system loading, distances to loads, and other design considerations. Two or three voltage levels are not uncommon in distribution systems as determined by the distances from transmission level drop points and specific regional loads.

This section will deal with the basics of transmitting and distributing electric power and is intended to provide an appreciation for some of the routine problems faced by power companies in delivering economic and reliable power to their customers.

3.1 TRANSFORMERS

An important byproduct of Faraday's discovery of magnetic induction was the transformer. In its simplest form, a transformer can be constructed by winding two independent insulated coils of wire around an iron rod, as illustrated in Figure 3.1-1 (▼).

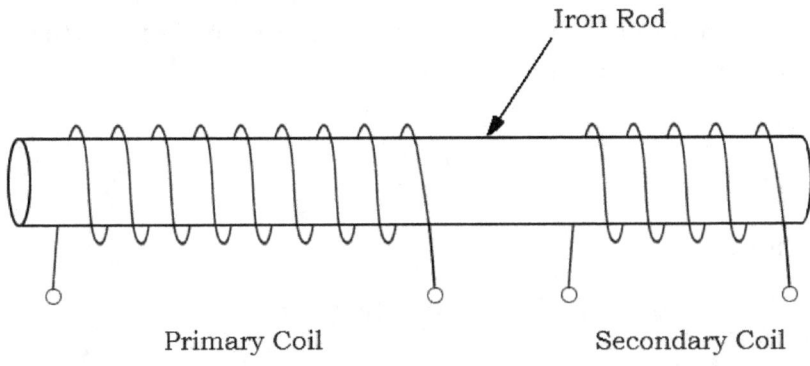

**Figure 3.1-1
Simple Transformer**

Faraday demonstrated that current flowing in one of the coils induces magnetic lines of flux in the iron rod, inducing current flow from the rod into the second coil providing the current flow is *changing*. The coil receiving driving current is designated the primary coil or winding and the other, the secondary coil or winding.

The application of alternating current into the primary winding induces a *changing* flux that satisfies the requirement that a *changing* current is necessary in transferring current into the secondary winding. This performance is consistent with Faraday's theory of induction, i.e., that induction from current carrying conductors to magnets and from magnets to conductors requires motion between the two.

The voltage induced into the secondary winding is proportional to the number of turns in each coil even though, neglecting small losses in the windings and transformer magnetic core, the level of electric power applied to the primary is transferred into the secondary. The relationship between voltage and transformer turns ratios is expressed as follows:

$$\frac{Np}{Ns} = \frac{Vp}{Vs}$$

where,

Np = number of turns on the primary

Ns = number of turns on the secondary

Vp = voltage applied to the primary windings

Vs = voltage at the secondary windings

Because transformers transfer power with small losses from their primary to secondary windings, the following expressions relating primary and secondary power levels may be written:

Electric Power Transmission And Distribution

$$VpAp = VsAs \quad \text{or}$$
$$\frac{Vp}{Vs} = \frac{As}{Ap}$$

where,

Vp = voltage applied to the primary windings

Vs = voltage at the secondary windings

Ap = primary current

As = secondary current

These phenomena provide the vital functions in electric power systems of raising or lowering voltages with very little power loss and lowering dangerous high voltages and currents to safe levels for measurement instrumentation.

The advantage of transmitting power at high voltage levels relates directly to Joule's A^2R or I^2R expression for power and the fact that as voltages are increased, power levels remain the same because less current is required to produce it. As we have seen, resistance, designated (R), is the only component of electrical loads that absorbs power; therefore, the inevitable I^2R losses caused by resistance in power lines will be reduced by the square of current decreases. Power line losses are often called "I^2R losses" as part of the industry's lexicon.

Assume a power line is transmitting 1 megawatt of power at 10,000 volts through a transmission line with 10 ohms of resistance. The current required is:

Electric Power System Fundamentals

$$P = VA = 10{,}000\,(A)$$

$$A = \frac{P}{V} = \frac{1 \times 10^6}{1 \times 10^4} = 100 \text{ amps}$$

where,

P = 1×10^6 watts = 1 megawatt, and

V = 10,000 volts

A = Current in amperes

The loss due to the A^2R effect is $(100)^2$ amps × 10 ohms or 100,000 watts or 10% of the transmitted power level.

$$\frac{1 \times 10^5 \text{ watts loss}}{1 \times 10^6 \text{ watts transmitted}} \times 100 = 10\%$$

Assume all of the parameters are the same except the transmission voltage is increased 10 times to 100,000 volts. In this case:

$$P = VA = 100{,}000\,(A)$$

$$A = \frac{P}{V} = \frac{1 \times 10^6}{1 \times 10^5} = 10 \text{ amps}$$

where,

P = 1×10^6 watts = 1 megawatt, and

V = 100,000 volts

A = Current in amperes

Electric Power Transmission And Distribution

The loss due to the A²R effect is (10)² amps × 10 ohms or 1,000 watts or 0.1% of the transmitted power level.

$$\frac{1 \times 10^3 \text{ watts loss}}{1 \times 10^4 \text{ watts transmitted}} \times 10 = 0.1\%$$

A 10 times increase in voltage, therefore, decreased losses by 100 times or the square of the voltage increase.

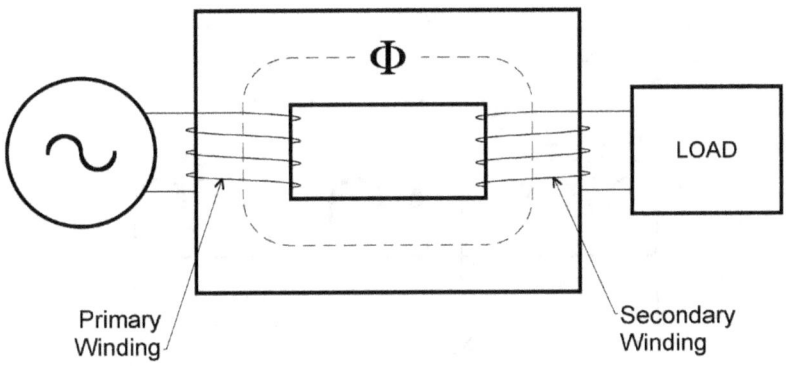

**Figure 3.1-2
Single Phase Transformer**

The transformer illustrated in Figure 3.1-2 (▲) is a single phase transformer with a closed iron core, allowing magnetic flux lines to rotate clockwise and counterclockwise as the primary winding current driver reverses in direction.

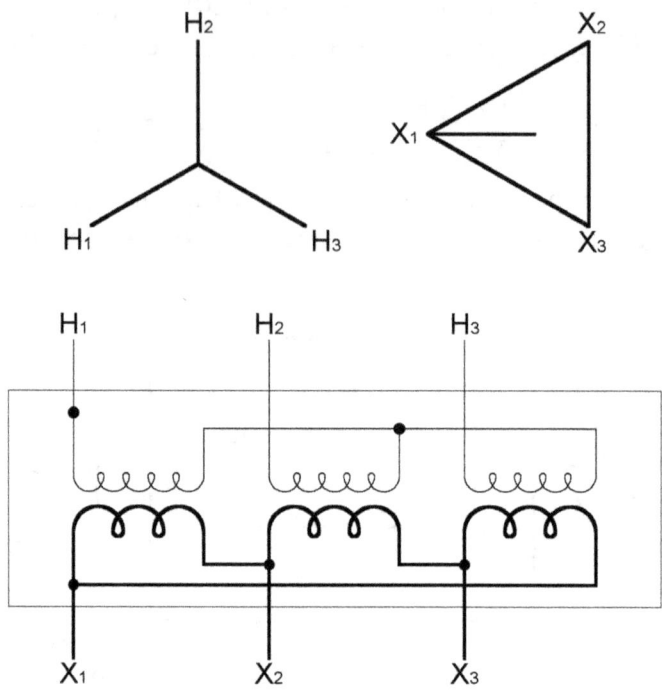

**Figure 3.1-3
Three Phase Transformer**

The three-phase transformer schematic illustrated in Figure 3.1-3 (▲) is electrically equivalent to three single-phase transformers built on a common core and mounted in a single case. Convention has established the higher voltage windings are designated as H_1, H_2, and H_3 and the lower voltage windings X_1, X_2, and X_3. The H_1, H_2, and H_3 windings are connected in what is termed a "Y" or "wye" configuration and the X_1, X_2, and X_3 windings in a "Δ" or "delta" configuration. It should be noted, a generator may be symbolized as a "Y" connected transformer without a primary winding.

Electric Power Transmission And Distribution

**Figure 3.1-4
High Voltage Power Transformer**

Electric Power System Fundamentals

The transformer pictured in Figure 3.1-4 (◄) is typical of a three-phase phase power transformer in the 50 MVA (mega-Volt-Ampere) size range. The three connections at the top right-hand side of the unit are the three-phase high voltage connections discernible because they connect through large insulators. The three connections at the top left-hand side of the unit are the three-phase low voltage connections as identified by their smaller insulators. These connections correlate with the "H" and "X" designations for high and low voltage windings. The unit illustrated is liquid-cooled as suggested by the external radiation cooling coils through which coolant is circulated after it is pumped through the internal windings of the transformer.

Transformers used in electric power systems are generally classified as either power transformers, or instrumentation transformers. Power transformers have turns ratios needed to raise specific voltage levels to higher or lower levels and range in physical size dependent upon their (VA) ratings.

Instrumentation transformers lower high voltages and currents to levels that are safe and practical for connection to measurement instrumentation. Voltage and current measurements provide data for calculating power factor, VARs, and for determining over voltage and over current situations.

Transformers that measure current are called *Current Transformers* and are designated as CTs throughout the industry. They have few turns on their primary windings and many turns on their secondary windings to take advantage of the expressions previously discussed:

$$\frac{Is}{Ip} = \frac{Np}{Ns} \quad \text{or} \quad Is = \frac{NpIp}{Ns}$$

where,

Np = number of turns on the primary

Ns = number of turns on the secondary

Ip = primary current

Is = secondary current

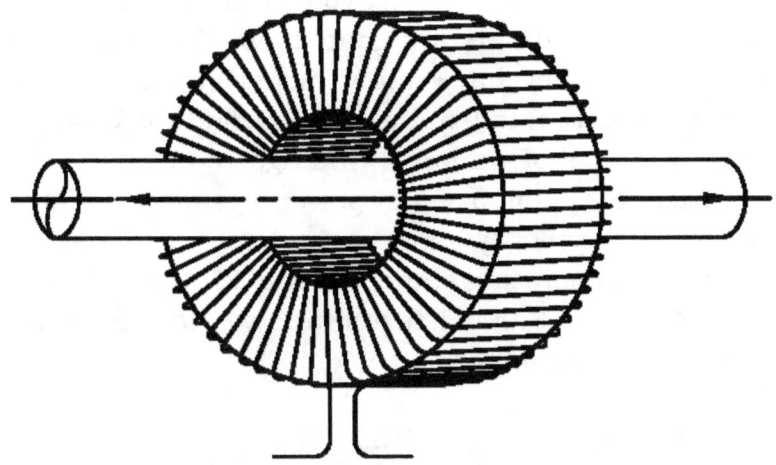

**Figure 3.1-5
Typical Current Transformer Construction**

Pictured in Figure 3.1-5 (▲) is a current transformer with a single turn primary and many turns on its secondary.

In a power line carrying 500 amperes, a CT with a 1:100 turns ratio would reduce that current to 5 amperes, i.e.:

Electric Power System Fundamentals

$$\text{Is} = \frac{\text{NpIp}}{\text{Ns}} = \frac{(1)(500)}{100} = 5 \text{ amperes}$$

where,

Np = number of turns on the primary

Ns = number of turns on the secondary

Ip = primary current

Is = secondary current

Transformers that measure voltage are called *Potential Transformers* and are designated as PTs throughout the industry. They have many turns on their primary windings and a small number of secondary windings. A large number of primary windings reduces the load or burden they impose on power systems. Potential transformers typically have turns ratios of 3000:1 or higher. In a generating station with a 13.8 kV generator bus, a PT with a 3,000:1 turns ratio would reduce 13.8 kV to 4.6 volts as follows:

$$\frac{\text{Np}}{\text{Ns}} = \frac{\text{Vp}}{\text{Vs}} \qquad \text{Vs} = \frac{(13,800)(1)}{3000} = 4.6 \text{ volts}$$

where,

Np = number of turns on the primary

Ns = number of turns on the secondary

Vp = voltage applied to the primary windings

Vs = voltage at the secondary windings

Electric Power Transmission And Distribution

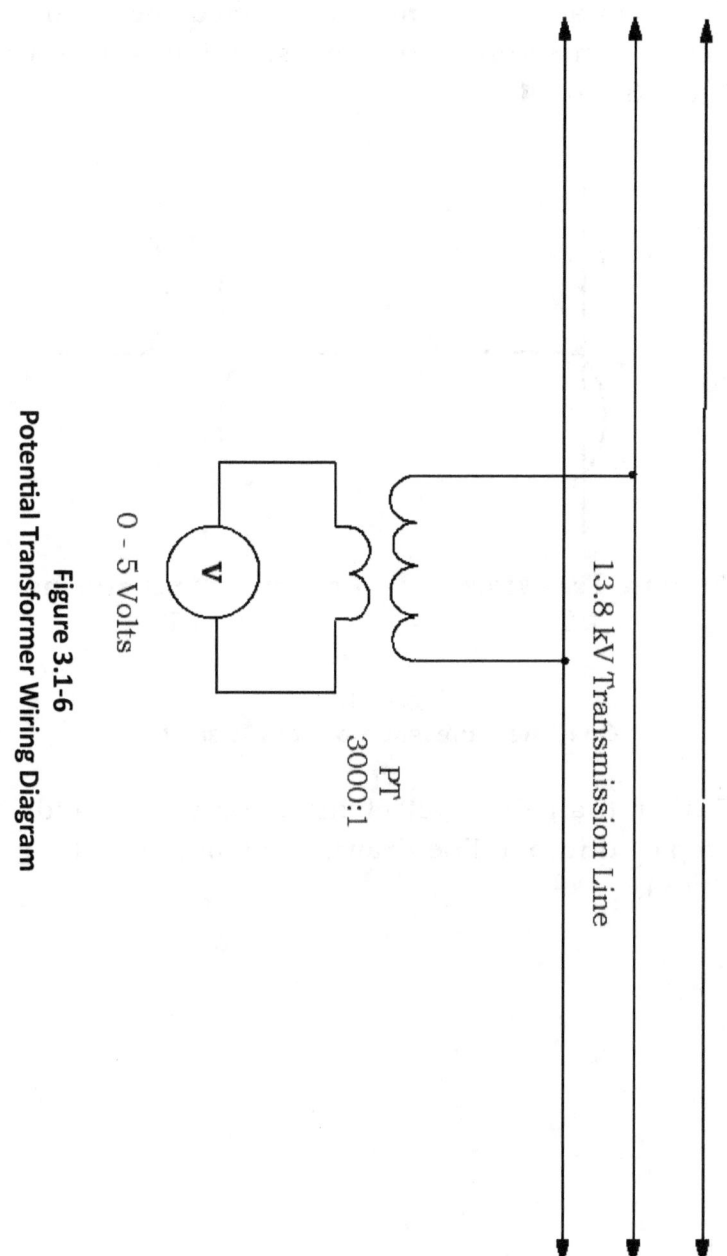

**Figure 3.1-6
Potential Transformer Wiring Diagram**

Electric Power System Fundamentals

A potential transformer connected between two phases of a three-phase transmission line is diagramed in Figure 3.1-6 (◄).

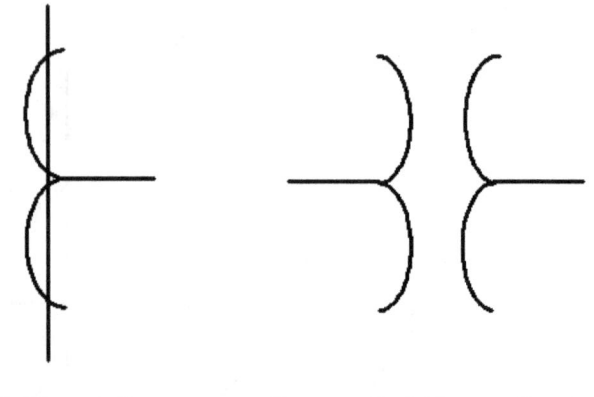

Current Transformer　　**Potential Transformer**
　　　CT　　　　　　　　　　　**PT**

Figure 3.1-7
One-Line representations of PTs and CTs

Rather than draw all of the details, CTs and PTs are depicted in one-line drawings as illustrated Figure 3.1-7 (▲).

Electric Power Transmission And Distribution

115 kV CT in a Distriution Substation

Electric Power System Fundamentals

CTs Measuring Low Voltage Three-Phase Current

Electric Power Transmission And Distribution

High Voltage Current Transformers

Electric Power System Fundamentals

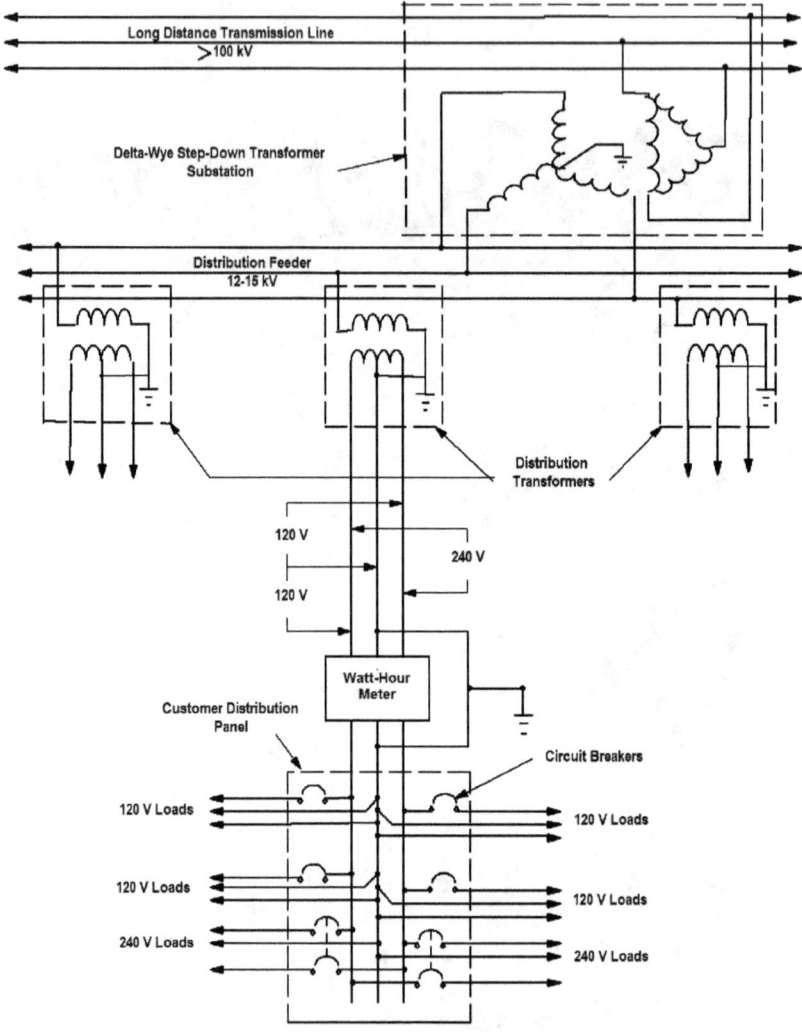

**Figure 3.2-1
Three-Line Diagram Transmission Level
to Distribution Feeder to Typical Residential
Distribution Panel**

Electric Power Transmission And Distribution

3.2 TRANSMISSION AND DISTRIBUTION WIRING

Electric power system drawings use symbols to illustrate functional components and to show how they are interconnected.

Figure 3.2-1 (◄) is a wiring diagram of the connections between a typical residential distribution panel and a 115 kV cross-country transmission line.

Beginning at the 115 kV transmission line, a delta-to-wye connected transformer lowers the high voltage to a 12-15kV distribution feeder. The distribution feeder is connected to several smaller 12-15kV-to-240 V distribution transformers. Each of the smaller transformers supplies a residential user with a center-tapped 240 V power connection. The center tapped transformer is used to supply appliances that require 240 V or 120 V. Loads from the distribution panel are supplied through current limiting protective circuit breakers. A watt-hour meter measures the power used by each customer who is billed according to usage. Note that each of the low voltage distribution transformers is connected to a different leg of the three-phase distribution feeder to balance the loads on each phase.

This is called a three-line drawing because it illustrates all three of the phase connections from the three-phase transmission line. It does not show the current and voltage measurement equipment installed between the high voltage transformer and the distribution feeder because the scale of the drawing would cause difficulty reading the added detail. Measurement of power supplied to the distribution feeder would be necessary to allow the transmission company to bill the distribution company.

Electric Power System Fundamentals

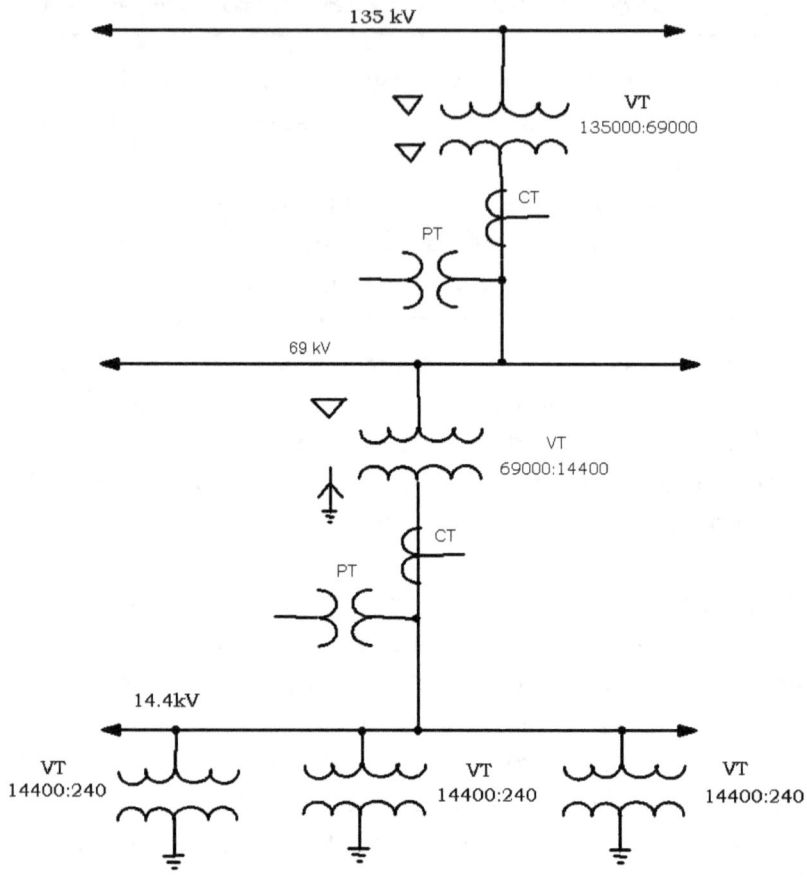

Figure 3.2-2
One-Line Diagram of HV Transmission Line to Intermediate Level to Distribution Feeder to Distribution Feeder to User Transformers

Figure 3.2-2 (▲) is a one-line drawing of the equipment illustrated on the three line drawing in Figure 3.2-1 with PT and CT instrument transformers and an intermediate 69 kV distribution line added. One-line

Electric Power Transmission And Distribution

drawings illustrate more equipment in less space when less detail is needed.

From the top down, a 135 kV transmission line connects to a delta-delta wound 135 kV to 69 kV transformer lowering the transmission voltage to an intermediate 69 kV distribution level. Power supplied to the 69 kV line is measured with PTs and CTs allowing the transmission line company to bill the distribution company for the power it uses.

A delta-wye 69 kV to 14.4 kV transformer lowers the intermediate distribution voltage to a local distribution feeder. Single phase 14.4 kV to 240 volt transformers supply end users whose billing is calculated from individual watt-hour meters.

Of necessity, energy management centers (EMCs) use even more abbreviated drawings to depict their control and monitoring operations. These diagrams are intuitively easy to understand. Examples of EMC drawings are used in grid network discussions in Section 3.3.

Photographs on the following pages illustrate a typical local distribution transformer and a residential distribution panel with circuit breakers.

The high voltage is connected to the large insulator on the top of the transformer case. The three connections to the center tapped secondary are seen on the side of the unit near the top.

Electric Power System Fundamentals

Distribution Transformer 14.4-to-240 V Center Tapped

Residential Distribution Panel 240/120 V

3.3 SUBSTATION FUNCTIONS AND CONTROLS

**Figure 3.3-1
A Typical Electric Power Distribution Substation**

Electric power system substations such as the one illustrated in Figures 3.3-1 (▲) are located at points within power systems where transmission voltage levels are dropped through transformers and connected to local distribution lines or elevated for long distance transmission. High voltage switches are provided in substations to isolate sections of a power grid that may be disabled or out of service, to provide optional routing of power, and to prevent damage to other connected equipment when faults are detected.

Instrumentation transformers are also provided in substations to measure voltages and currents, allowing

Electric Power System Fundamentals

the calculation of power factor and VAR values and to determine the direction in which power is flowing throughout a power system.

When high voltage switch contacts that are under load are pulled apart, arcing occurs between the contacts. Arcing damages switch contacts and must be suppressed to extend the number of open-close cycles the switch can endure before replacement or maintenance is necessary. There are four primary types of high voltage switches: those that use air, oil, gas, or a vacuum, for arc suppression. Each type has advantages and disadvantages in cost, durability, and maintainability. High voltage switches are referred to as circuit breakers and noted on drawings as ACBs, OCBs, Gas or Vacuum breakers.

**Figure 3.3-2
Typical 3-Phase High Voltage Switch**

Electric Power Transmission And Distribution

Switch gear in power systems is treated as a specialized category of equipment in terms of how switching at various voltage and power levels is controlled and implemented.

Substations are often named for their specific function such as a "switching" station, or "step-up" station, or a "distribution" station. The one-line drawing in Figure 3.3-3 (▶) depicts the major equipment within a typical distribution substation. Switching and instrumentation are not included in the drawing, only the connectivity within the substation is shown. This substation is fed by transmission lines from two generating stations, has connections to two other transmission lines, and includes two step-down transformers that supply independent distribution feeders. The dark line is an electrical bus. Electrical buses are connection points along which the voltage is constant in both magnitude and phase angle. Electrical buses provide a common point for line connections within a substation.

Switching of high voltage within substations can be configured in various ways depending upon performance criteria, reliability requirements, failure mode evaluations, cost, and other factors. The substation one-line drawing in Figure 3.3-4 (▶) illustrates the substation of Figure 3.3-3 (▶) including circuit breakers. The configuration shown is called a "single bus arrangement." In this configuration either of the high voltage lines or step-down transformer connections can be opened or closed independently.

Electric Power System Fundamentals

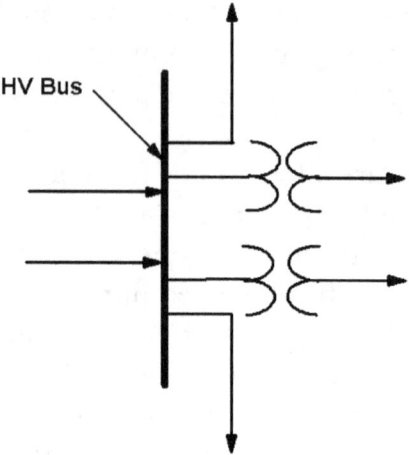

**Figure 3.3-3
Distribution Substation One-Line**

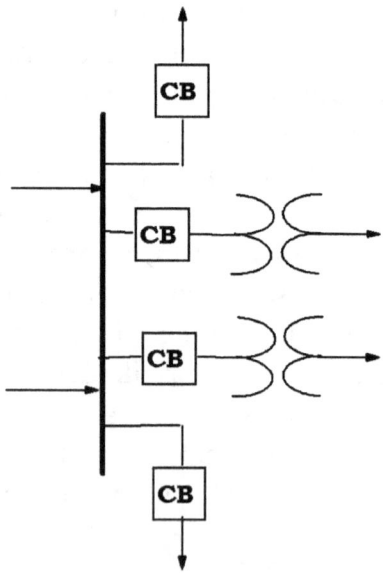

**Figure 3.3-4
Distribution Substation One-Line**

Electric Power Transmission And Distribution

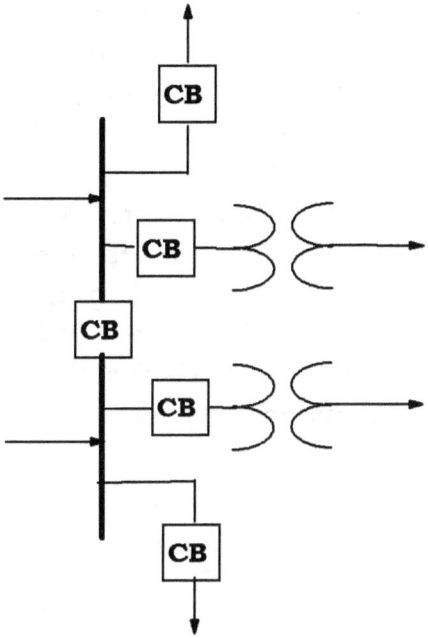

**Figure 3.3-5
Distribution Substation One-Line**

If a circuit breaker is added to allow isolation of the bus into two parts, it is called a "single bus with tie breaker" configuration as illustrated in Figure 3.3-5 (▲).

In addition to the single bus and single bus with tie breaker switch configurations, four other configurations have become "standardized." They are listed below in order of increasing complexity:

1. The ring bus,
2. The main and transfer bus,
3. The breaker and a half, and
4. The double bus-double breaker.

Substation configuration choices depend upon such factors as reliability analyses, flexibility, and cost.

The connectivity and measurement data from substations are essential to successful power grid management.

It should be noted that electric power substations are isolated behind secure barriers for safety reasons preventing individuals from entering and becoming exposed to dangerous high voltage levels. Only qualified personnel should be permitted within the safe boundaries surrounding substations.

Electric Power Transmission And Distribution

Sections 3.0 Questions:

1. What voltage level is considered the dividing line between transmission and distribution voltages?

2. What are some of the factors that determine the selection of transmission and distribution voltage levels?

3. How do the primary and secondary turns in a transformer relate to its primary and secondary voltages?

4. Why is power transmission efficiency improved by raising the voltage level in a transmission line?

5. What functions are performed by instrumentation transformers?

6. What is the difference between PT and VT transformers?

7. What are the advantages and disadvantages of one-line and three-line wiring diagrams?

8. Why are three-phase transformers designated as wye or delta types?

9. Why is less detail shown on EMC level drawings?

10. Why are single-phase center-tapped transformers used on residential distribution transformers?

Electric Power System Fundamentals

11. Where is typical residential end-user power measured and with what type device?

12. Name three purposes for which high voltage switches are used in power system substations.

Notes

Notes

4

ELECTRIC POWER GRID MANAGEMENT

The management of electric power networks is a large subject that could easily expand beyond the limits established for this discussion. This discussion is intended to provide a fundamental understanding of how power flow within networks is measured and how the information is used to provide the very high reliability maintained throughout the electric power industry that is so often taken for granted.

4.1 GRID POWER FLOW

The measurement of power flowing between various points in a transmission and distribution system is needed to determine when lines are overloaded, could accept more power if contingencies occurred, and to provide the most cost effective routing of power. Power flow is also important in determining exchange between entities for billing purposes and to determine when to consider

Electric Power System Fundamentals

expansion or design of new systems or components of systems.

Figure 4.1-1 (▶) is a one-line diagram of a three-phase transmission line of impedance Z with connecting buses at each end. As previously stated, an electrical bus is a connection point along which the voltage is constant in both magnitude and phase angle. The voltage from generators G1 and G2 are stepped-up through transformers T1 and T4 to the desired transmission voltage level and applied to the left-hand bus. The left-hand bus is then connected through a transmission line to the right-hand bus. The voltage from generator G3 is stepped-up through transformer T7 to the transmission voltage and applied to right-hand bus. Although the voltages at each end of the transmission line are the same amplitude, power flows between them because the voltage at the left-hand bus leads the voltage at the right-hand bus by $12°$ as indicated by the angular notations at the top of the right-hand bus.

Electric Power Grid Management

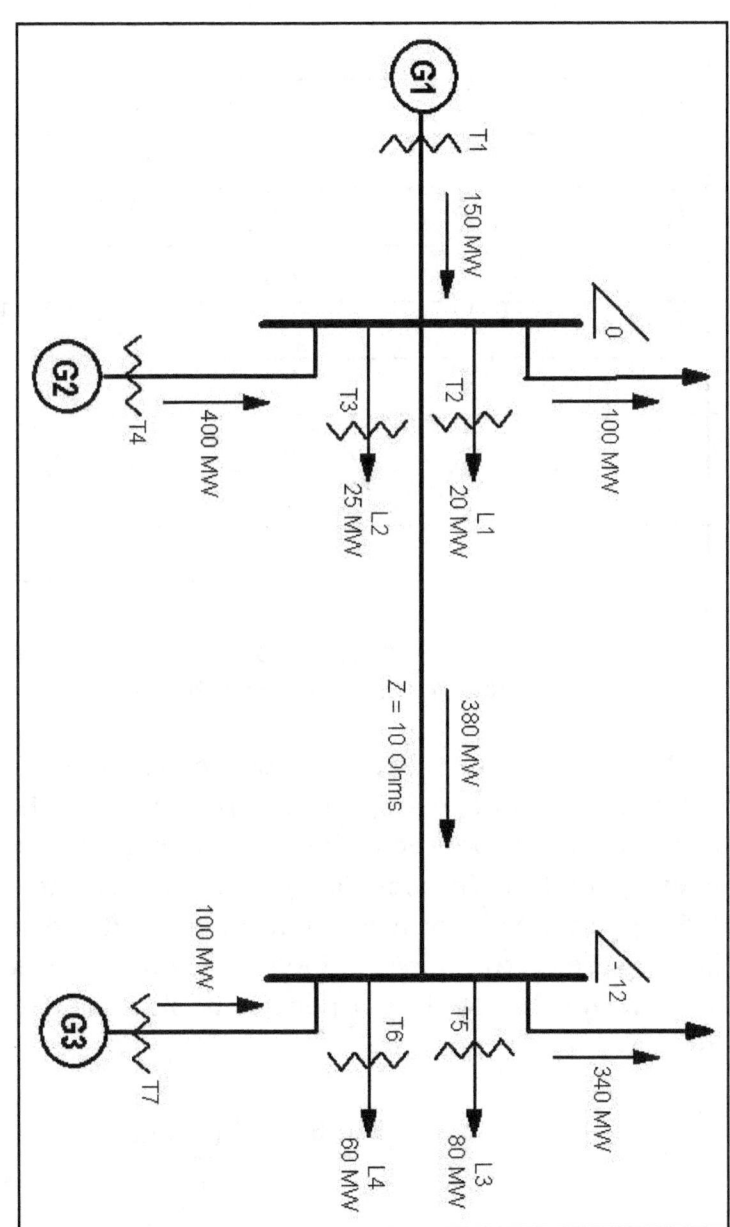

Figure 4.1-1
One-Line Diagram of a Transmission Line Between two Busses

Electric Power System Fundamentals

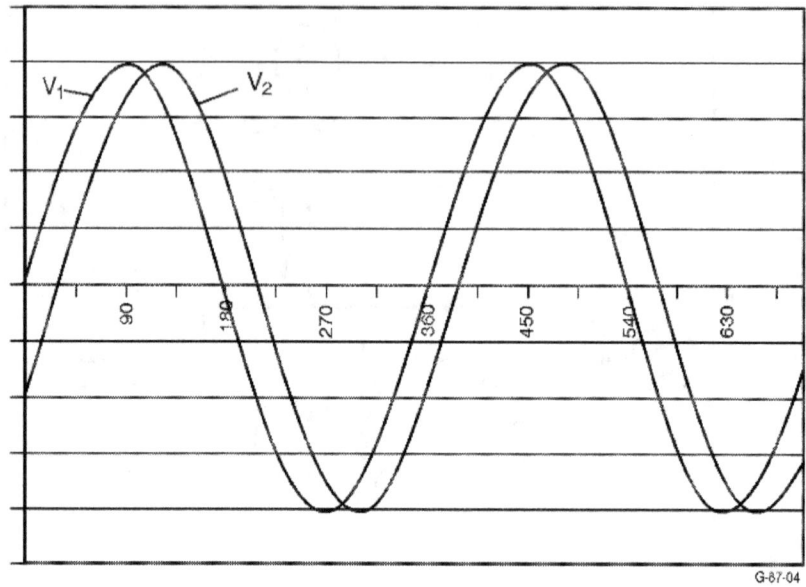

**Figure 4.1-2
Voltage Off-Set between Points**

A typical voltage offset is illustrated in Figure 4.1-2 (▲). Although the voltage amplitudes are the same, the offset or phase shift between the two causes the voltages at every instant in time to be different. If the voltage difference caused by the angular phase shift is expressed as the sine of the offset angle, the power flow between two buses connected by a power line may be expressed as:

$$P = \frac{V_1 V_2 \sin\theta}{Z_L} \approx \frac{V^2(\sin\theta)}{Z_L}$$

108

where,

  P = Power in watts
  V = Voltage in volts
  A = Current in amperes
  Z_L = Line impedance in ohms
  θ = Offset angle in degrees

For example, if the line impedance between two busses is 10 ohms, and the voltage offset is 12° at 138kV, the power flowing through the transmission line connecting the two busses is calculated as follows:

$$(138 \times 10^3)^2 (0.20)/10 = 380 \text{ MW}$$

where,

  0.2 = Sin of 12°

The direction of power flow will be away from the leading voltage source.

This example also illustrates why any generator contributing power to an infinite grid increases its power contribution when torque applied to its drive shaft is increased, thus driving its phase angle "ahead" of other generators producing power into the grid even though its voltage amplitude is unchanged.

Electric Power System Fundamentals

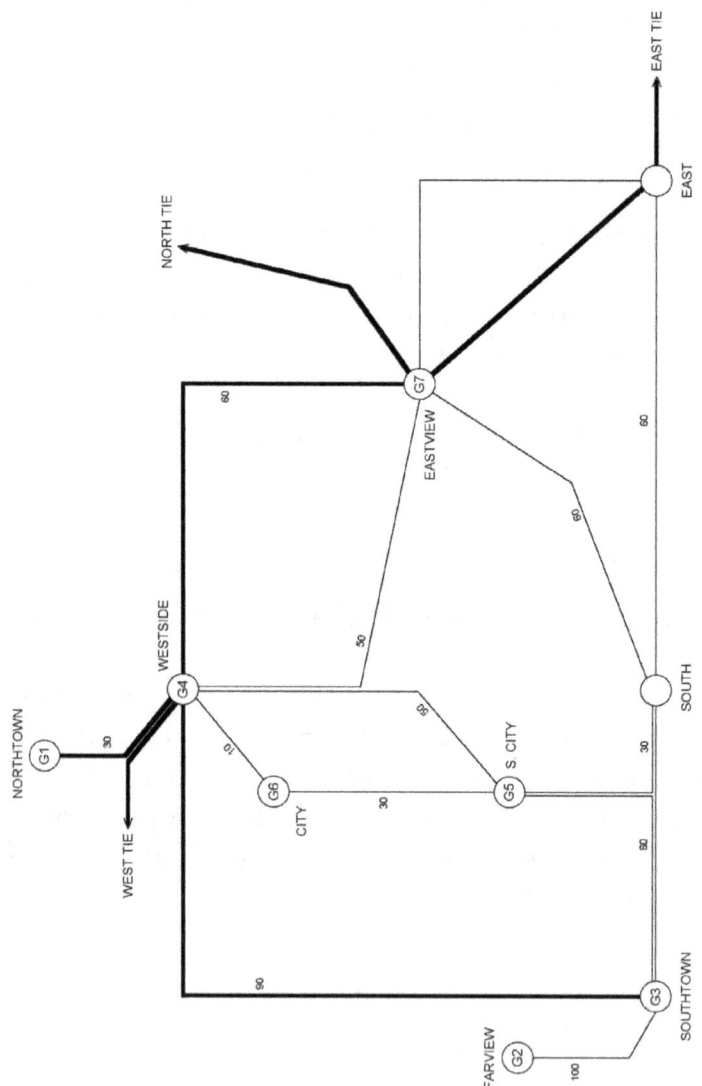

**Figure 4.1-3
Geographical Layout of a Typical Power System Segment**

Electric Power Grid Management

Figure 4.1-3 (◄) illustrates the geographical locations of the major components in part of a typical power system. It includes seven (7) power generation plants, and ties to the North, East, and West. The transmission line distances are given in miles.

To allow a power flow study of this system, it is necessary to construct a system one-line drawing illustrating the generators, buses and tie points. The connectivity within the system is subject to change, in which case, the drawing is re-constructed. The connectivity is often referred-to as the topography of the system. System topography is derived from circuit switch position information transmitted to the EMS from substations throughout the grid. The resulting diagram is called a system load flow or a base case for load flow study and is illustrated in Figure 4.1-4 (►).

Figure 4.1-5 (►) is called a "System Load Flow Diagram," which indicates the power flowing in each branch. The section within the dotted line is at 138 kV and outside 345 kV.

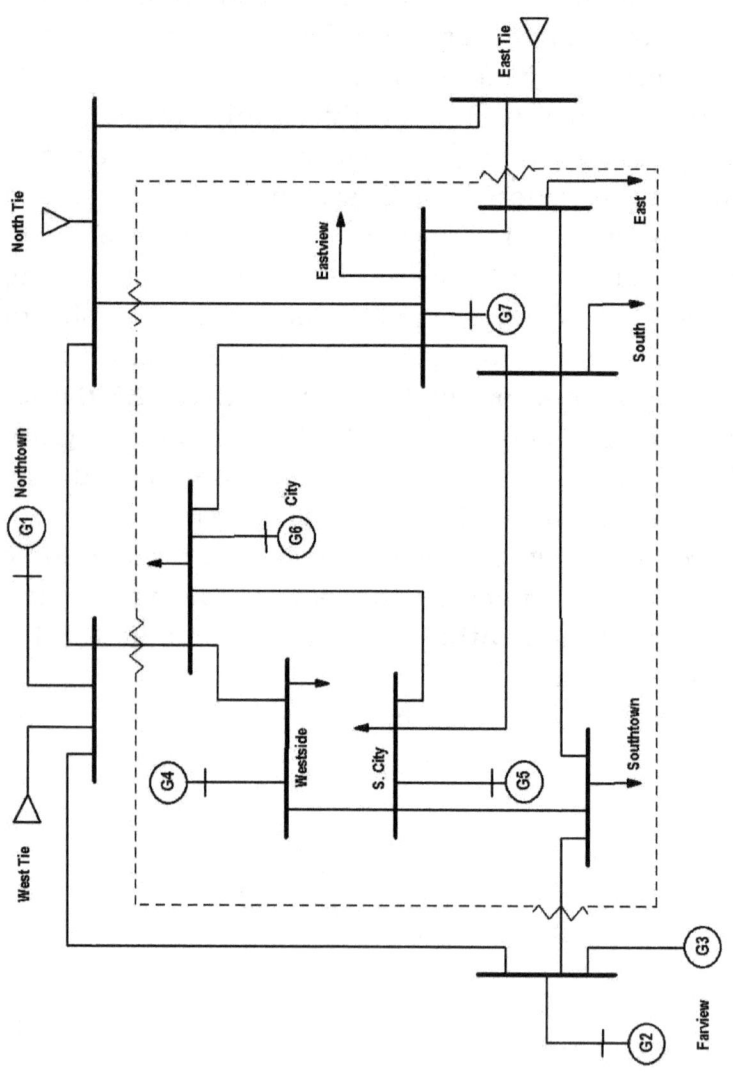

Figure 4.1-4
Base Case for Load Flow Study System One-Line Diagram

Electric Power Grid Management

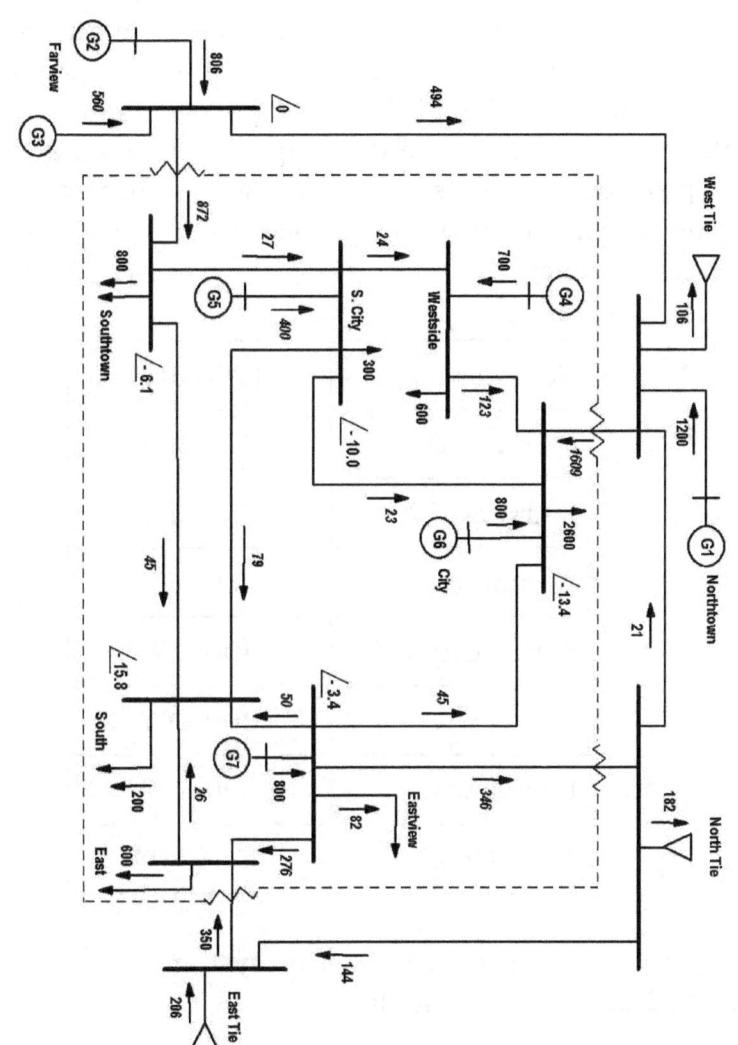

**Figure 4.1-5
System Load Flow**

113

Electric Power System Fundamentals

Phase angles throughout the system are referenced to the Fairview 345 kV bus. The Fairview bus was chosen arbitrarily and is referred-to as the "swing bus." The MW power flow is given for each branch with an arrow indicating direction. The total load served from each substation and the 345kV and 138 kV transformers are shown. To simplify the drawing, the generating station step-up transformers are not shown. To obtain the total generation, the power supplied from each station is added as follows:

Northtown G1	1200 MW
Fairview G2	806 MW
Fairview G3	560 MW
Westside G4	700 MW
South City G5	400 MW
City G6	800 MW
Eastview G7	800 MW
Total Gen.	5266 MW

Note: Power losses within the system are not shown including parasitic load losses.

To obtain net interchange, the directional megawatt flow at each tie is summed as follows:

Electric Power Grid Management

West tie	+106 MW
North tie	+182 MW
East tie	− 206 MW
Net Interchange	+ 82 MW

The total load is:

Total Generation	5266 MW
Net Interchange	82 MW
Total Load	5184 MW

(The total load within the defined system includes losses.)

4.2 CONTINGENCY ANALYSIS

One of the principal reasons for conducting and maintaining system topography and load flow studies is to provide information for the analysis of what-if scenarios should contingencies occur in the operation of a power grid.

Contingency analysis answers such questions as: If a particular transmission line trips out of service, will the adjacent lines or paths of power flow be overloaded, or if a power interchange transaction is allowed, how will the added power be redistributed throughout the system and will overloads occur?

Contingencies are very important in judging the

overall adequacy of a system configuration in meeting its load demand without violating equipment ratings or its security. A system is said to be secure if a single outage will not cause further outages. An insecure system can cause a cascading failure leading to a total blackout.

A contingency involving a single change may be predictably easy to deal with; however, when more than one change occurs, predictability becomes complicated. Contingency analysis software is capable of running a series of load flows that would result from various contingencies. An operator can then use these hypothetical load flows in several ways. For example, if a transmission line is operating near its MVA limit and is in danger of overload should another contingency occur, the operator could shift some of the load on the endangered line to another generation plant or adjust interchange outside of the area.

For the purpose of illustration, one such contingency is presented in Figure 4.2-6 (▶). In this case, the 138 kV line between the South substation and the South City generation station has been taken out of service as illustrated by the dashed line

The other three lines into the South Substation have picked-up the 79 MW that was flowing in the removed 139 kV line. The Eastview and Farview generating plants have added 162 MW to maintain the same flows at the West, North and East ties and Southtown, South, and East loads. The 1 MW discrepancy at the Southtown bus and 1 MW discrepancy at the Eastview bus are caused by losses in the transmission lines from the added power flow.

Electric Power Grid Management

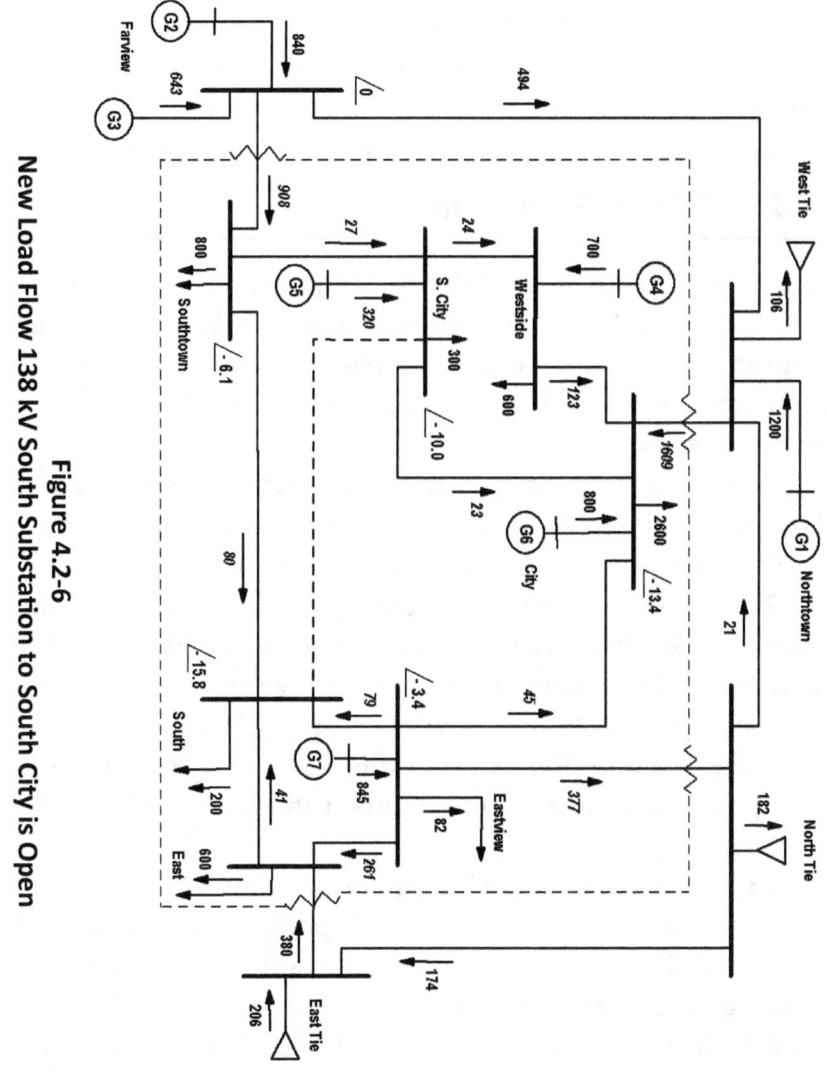

Figure 4.2-6
New Load Flow 138 kV South Substation to South City is Open

117

Although the power system remains stable in this illustration with one line outage, it is conceivable that a second outage could overload lines such as the 138 kV line between the Southtown and south substations. The capability of generating stations may also be unable to sustain added loads.

4.3 LOAD PREDICTION

Load prediction provides an estimate of how much generation will be needed in advance of actual load demand allowing time to make critical cost impact decisions to start, regulate, or to temporarily shutdown power plants.

The history of daily power demand is an important tool in predicting future loads. Other contingencies such as weather forecasting are also important.

Figure 4.3-1 (▶) illustrates a typical weekly load curve indicating the historic normal loads for each day and hour. The lower area of the curve is the minimum continuous load or base load that must always be served. The intermediate load is the daily periodic variation, and the peak area represents loads that exceed the daily average.

In scheduling generation, some amount of reserve is necessary. Typically 15-to-20 percent is considered prudent; however, it must be "spinning." Spinning reserve is generation that is on-line or generation that can be brought on-line within a short time. Generation that can be on-line within 10 minutes is often referred-to as spinning reserve.

Electric Power Grid Management

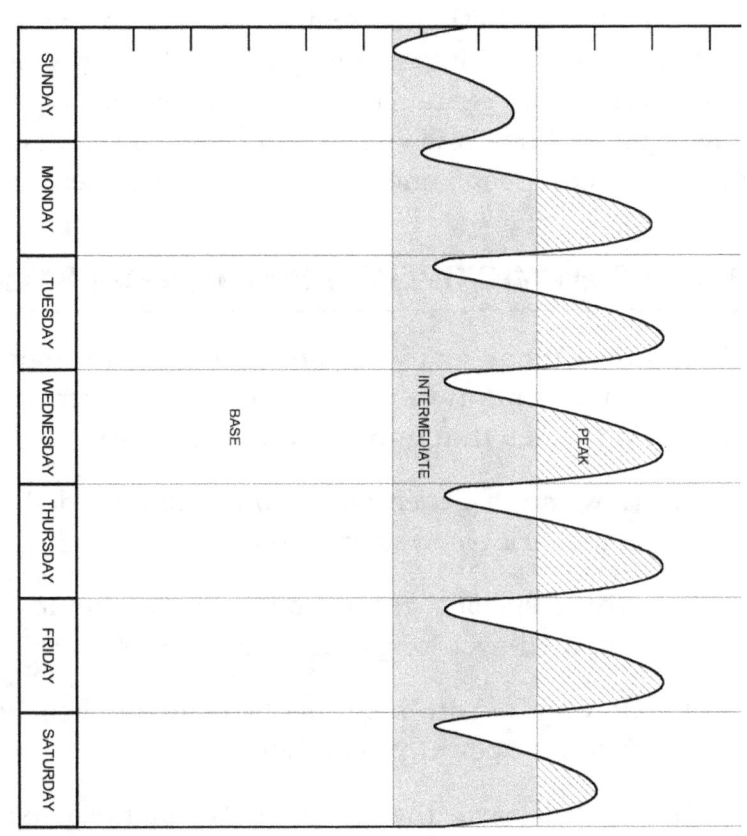

**Figure 4.3-1
Weekly Load Curve**

It is necessary to start-up small power generating stations called peaking units when short-term load peaks are encountered. Peaking units may not provide the most efficient power but are less expensive than starting a larger plant or building more power capability into a power system.

Links to the National Weather Service in the United States is a very important part of load prediction. Thunderstorms, tornadoes, and hurricanes can take down power lines and disrupt normal power flow in a system as can floods and other natural disasters.

4.4 ECONOMIC OPERATIONAL CONSIDERATIONS

There are a great many actions that system operators can exercise to improve the economy of a power system. The major issues that impact system cost are:

1. How much generation should be provided to meet variable system load demands?

2. Which specific generation stations should be operated over long and short terms?

3. How should on-line units be loaded to achieve optimum power production cost?

As we will see, the optimum economic dispatch of power occurs when the power delivery system is configured for optimum cost to deliver the power needed without contingencies that may cause unacceptable power outages or system instability.

Electric Power Grid Management

4.4.1 ECONOMIC DISPATCH

Economic dispatch is the process of loading each on-line generation station for minimum cost. Minimum cost is achieved when further shifting loads among on-line units no longer lowers the overall production cost. That point occurs where the production costs of the on-line units crossover. The following two-unit example illustrates the process.

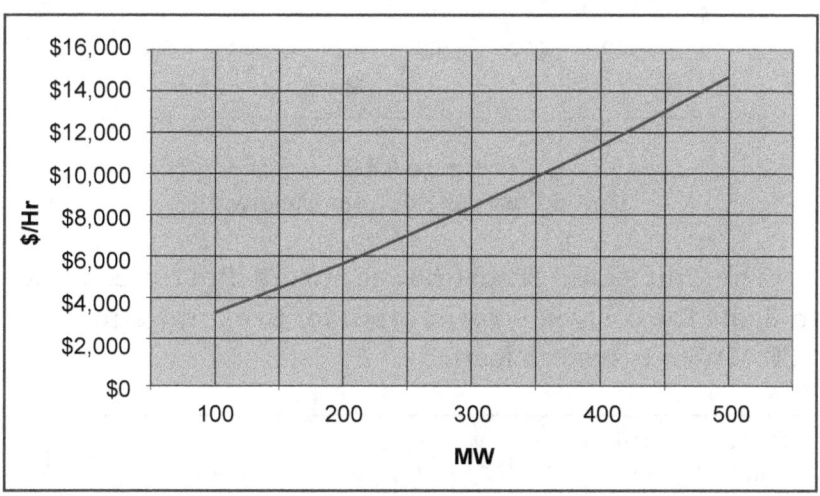

Figure 4.4-1
Unit A Cost per Hour per Megawatt

Electric Power System Fundamentals

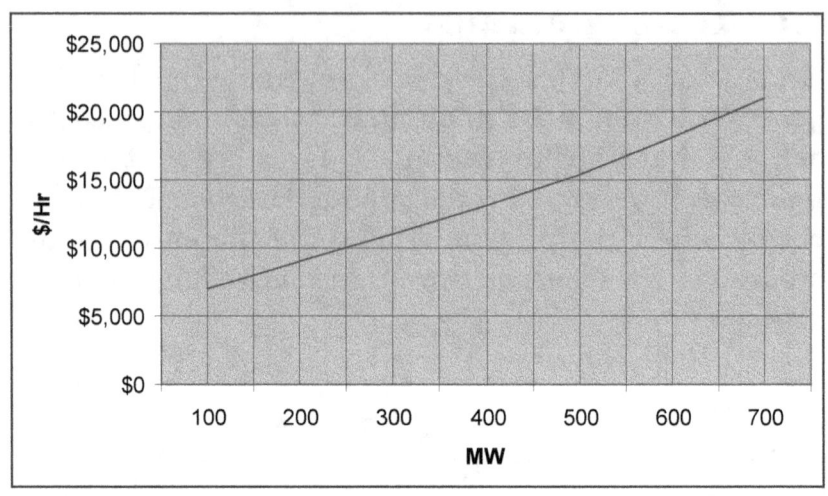

Figure 4.4-2
Unit B Cost per Hour per Megawatt

The Unit A and B cost power curves illustrated above indicate the cost vs. power output for two units supplying 800 MW to a system load.

Unit A Load MW	Unit B Load MW	Units A + B MW	Unit A $/Hr	Unit B $/Hr	Total Cost $/Hr
100	700	800	$3,300	$21,000	$24,300
200	600	800	$5,700	$18,100	$23,800
300	500	800	$8,400	$15,400	$23,800
400	400	800	$11,400	$13,100	$24,500
500	300	800	$14,700	$11,000	$25,700

Table 4

Table 4 (◄) illustrates the steps in an iterative process to achieve the minimum cost from the power curves. At step (1) in the iterative process, unit (A) indicates a cost of $3,300/hr. when producing 100 MW. Unit (B) indicates a cost of $21,000/hr when producing 700 MW. Together the two units are producing 800 MW at a total cost of $24,300/hr.

At step (2) unit (A) generation is increased to 200 MW and unit (B) generation is reduced to 600 MW to produce the desired 800 MW. In this case the cost curves show an increase in unit (A) cost of (+) $2,400 and a decrease in Unit (B) cost of (−) $2,900.

At step (3) unit (A) generation is further increased to 300 MW and unit (B) is further lowered to 500 MW. The 800 MW total remains the same; however, in this case the unit (A) incremental increase is offset by the unit (B) incremental decrease resulting in a net change of zero dollars.

To complete the example, step (4) shows an increasing cost if the process of increasing the output of unit (A) and decreasing the output of unit (B) is continued.

Computer programs eliminate manual calculations such as those used in this example and produce the optimum unit dispatch information regardless of the number of on-line units.

4.4.2 UNIT COMMITMENT

The determination of which generation stations to place on-line and over what time interval is called, "unit commitment." The objective in unit commitment is to develop a generation schedule for the anticipated load

Electric Power System Fundamentals

demand plus reserve without committing more units than necessary. The generation efficiency of each unit is also an important consideration in unit commitment scheduling.

Unit commitment includes other considerations. For example, the load requested from a particular plant may be below or above its maximum heat rate efficiency or it may have start-up and shutdown time limitations or may be impacted by the number of heating and cooling cycles that are imposed with a consequent impact upon maintenance.

Electric Power Grid Management

Section 4.0 Questions:

1. If the voltage amplitude is the same at each end of a transmission line, why does power flow through the line?
2. How is power flowing between two points on a transmission line measured?
3. What is meant by "net interchange" in a power system?
4. What is the difference between a system one-line drawing and a system power flow drawing?
5. Why is contingency analysis important?
6. What is the difference between net interchange and total generation?
7. What is the purpose of economic dispatch?
8. What is the difference between base loads intermediate loads and peak loads in power systems?
9. What is the difference between economic dispatch and unit commitment?

Notes

APPENDICES

Appendix A
BATTERY CHEMISTRY

Battery Chemistry

Electromotive forces are created within batteries by chemical processes called oxidation and reduction. These processes take place because of electrically charged particles called ions* that travel within the battery between the positive and negative battery terminals. An explanation of oxidation and reduction processes and the formation of ions requires a review of atomic structures and chemical notations.

Atoms are comprised of a nucleus of protons and neutrons that form the greater part of its mass, and electrons that revolve around the proton/neutron nucleus. Protons are positively charged, neutrons have no charge, and electrons are negatively charged particles. In neutral atoms, the number of negatively charged electrons equals the number of positively charged protons. If a neutral atom gains or loses one or more electrons it becomes an ion* and the process is called ionization. In gaining electrons, atoms have a net negative charge and in losing electrons they have a net positive charge.

When writing the chemical expression for an ion its net charge is written in superscript. For example, a

---

* The word Ion derives from the Greek word iov that can be translated as "going."

zinc ion with two more protons than electrons has a net charge of +2 and is noted as:

$$Zn^{2+}$$

An oxygen ion with two more electrons than protons has a net 2 negative charge and is noted as:

$$O^{2-}$$

An ion with a net single negative or single positive charge is noted as:

$$Cl^- \text{ or } Na^+$$

Oxidation refers to the process of losing electrons. Reduction refers to the process of gaining electrons. In batteries, oxidation and reduction processes are simultaneous when its terminals are connected through an external conductor. The action is referred to as a redox process.

The figure below illustrates a typical voltaic cell consisting of a zinc electrode and a copper electrode separated by a porous barrier and immersed in a solution of a dissolved salt of the corresponding metal. These solutions that conduct ions are called electrolytes. The porous barrier prevents the two solutions from rapidly mixing but allows ions to diffuse through it.

The excess electrons that remain (on the Zn electrode) when Zn^{2+} ions emerge from the left-hand cell, cause it to become negatively charged. Likewise the gain of positive charges in the right-hand copper electrode when Cu^{2+} ions are absorbed, cause it to become positively charged.

Battery Chemistry

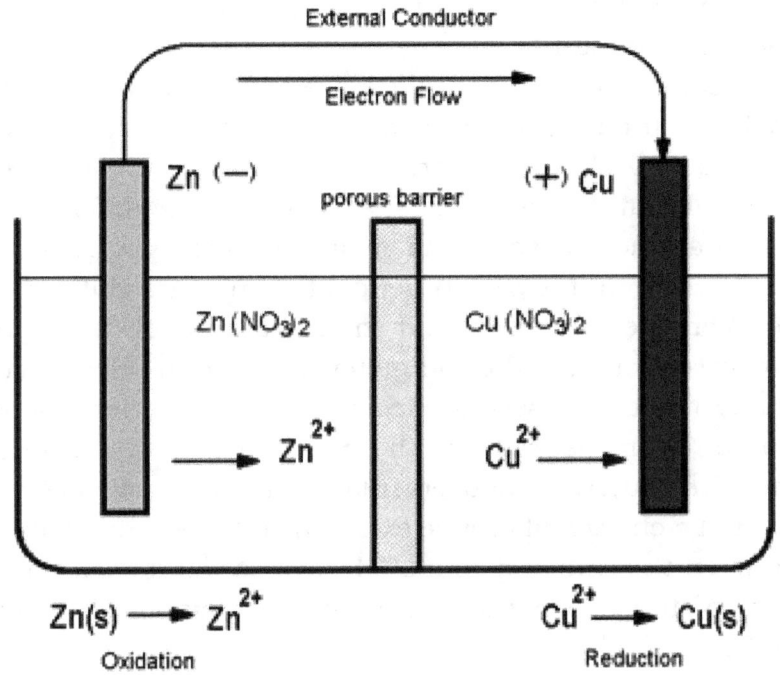

When an external connection is made between the zinc and copper electrodes, the negative charges in the zinc are attracted to the positively charged copper electrode. When that occurs, the net reaction in the oxidation of zinc and the reduction of copper is expressed as:

$$Zn(s) + Cu^{2+}(aq) \rightarrow Zn^{2+}(aq) + Cu(s)$$

where,

(s) = solid metal

(aq) = aqueous solution in which the solvent is water

Electrons flowing between the battery electrodes can perform useful electrical work, the rate of which is determined by the amount of electrical resistance in the connected circuit.

If the two halves of the cell are examined separately, each will produce a charge difference at their respective interfaces with the electrolyte in a very thin layer of only a few atomic diameters. All such charge differences are quantified in volts and measured with volt meters; however, even if it were practical to connect a volt meter between the thin layer and the metal of each half-cell, the introduction of the volt meter lead into the electrolyte would react as a second electrode and two interactions would be measured by the meter rather than one; therefore, such a measurement would be impractical even though a charge difference exists. A volt meter connected between the zinc and copper electrodes will measure the sum of the two metal-to-electrolyte interfaces in each half cell.

Batteries constructed from different elemental metals will produce different half-cell voltage potentials as indicated in the following table. Nineteen (19) elemental electrodes and their potentials compared to a standard hydrogen cell under standard laboratory conditions are given in the table. The potential in volts between the electrodes of a zinc-copper voltaic cell is the algebraic sum of the zinc potential (–0.758) and the copper potential (+0.344) or 1.102 Volts.

Battery Chemistry

Element	Potential In volts
Potassium	−2.922
Barium	−2.90
Sodium	−2.712
Magnesium	−2.40
Aluminum	−1.7
Manganese	−1.10
Zinc	−0.758
Chromium	−0.557
Iron	−0.44
Cadmium	−0.397
Nickel	−0.22
Tin	−0.13
Lead	−0.12
Hydrogen	0.00
Copper	+0.344
Silver	+0.799
Mercury	+0.86
Platinum	+0.863
Gold	+1.36

Appendix B

GENERATOR SYNCHRONIZATION

Electric Power System Fundamentals

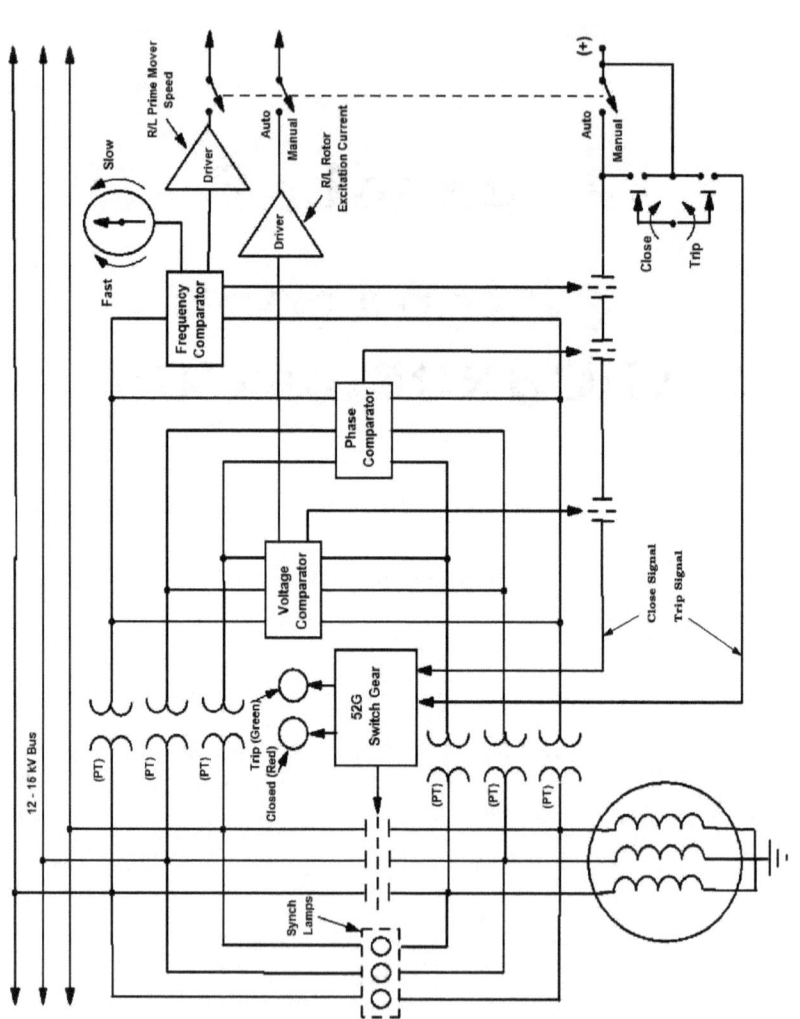

GENERATOR SYNCHRONIZATION

Connecting a generator to an already powered bus or to a power grid requires a synchronizing procedure. The purpose is to assure the generator's output matches the voltage, phase and frequency of the system to which it will be connected before actually making the connection. Generator connection circuit breakers are universally called 52G breakers.

The adjacent figure is a functional block diagram illustrating the circuitry needed to verify the parameters match before permitting closure of the 52G breaker.

Potential transformers are used to measure voltage from opposite sides of the 52G breaker for comparison of voltages frequencies and phases. Three comparators are used; one for each of the parameters that must be equal before the generator circuit breaker is closed. The comparators close independent sets of contacts such that the "Close" signal to the 52G switch gear occurs only if all three comparators issue a close signal. If the Automatic/Manual switch at the right-hand side of the diagram is in the "Auto" position when this happens the 52G breaker will close.

In the Automatic mode, the driver circuits labeled Raise/Lower (R/L) Prime Mover and Rotor excitation current drive their respective "set points" until the synchronizing conditions are reached. When operators select the manual mode of synchronization, the Raise/Lower signals must be manually controlled as follows:

1. Manually raise or lower the Rotor Excitation Driver set point as illustrated in Figure 2.2.5 until the metered value of the generator voltage equals the measured voltage of the 12-15 kV bus.

2. Manually raise or lower the Prime Mover set point in Figure 2.2.5 until the synchroscope in the upper right-hand section of the Synchronizing Functional Block Diagram rotates slowly in the clockwise (fast) direction. Clockwise rotation indicates the generator frequency or speed is greater (faster) than the 12-14 kV bus. Counterclockwise rotation indicates the generator power frequency is slower than that of the bus.

3. As the synchroscope reaches the 11 o'clock position, rotate the Trip/Close switch in the lower right section of the Synchronizing Functional Block Diagram clockwise to the "Close" position. If, at this time the three comparator contacts are closed, the 52G breaker switch gear will receive a "Close" command. The Trip/Close switch is a spring-loaded center-off switch that is operated momentarily and then returned to the center-off position.

Generator Synchronization

After the 52G breaker is closed, the "Trip" switch can be operated at any time to open the 52G breaker; however, it is not advisable unless an emergency situation arises or the generator load decreases to a low value. If the 52G breaker is opened under partial or full load, excessive arcing will occur damaging the breaker contacts. The sudden loss of load will result in a large transient signal to slow the prime mover, stressing steam controls or speed control limits in gas turbine prime movers.

The 52G switch gear illuminates a RED indicator lamp when the breaker is closed and a GREEN indicator lamp when open. A set of three "synch lamps," connected between the phases provide a redundant indication to assure operators of the proper voltage, frequency, and phase, match before closing the 52G breaker. The lamps indicate voltage differences between each of the phases and are brightest when the differences are greatest. They do not illuminate when synchronizing conditions are correct. Although they do not provide precision indications they do provide a meaningful indication if phase rotation is incorrect.

Appendix C

GENERATOR PROTECTIVE CIRCUITRY

Electric Power System Fundamentals

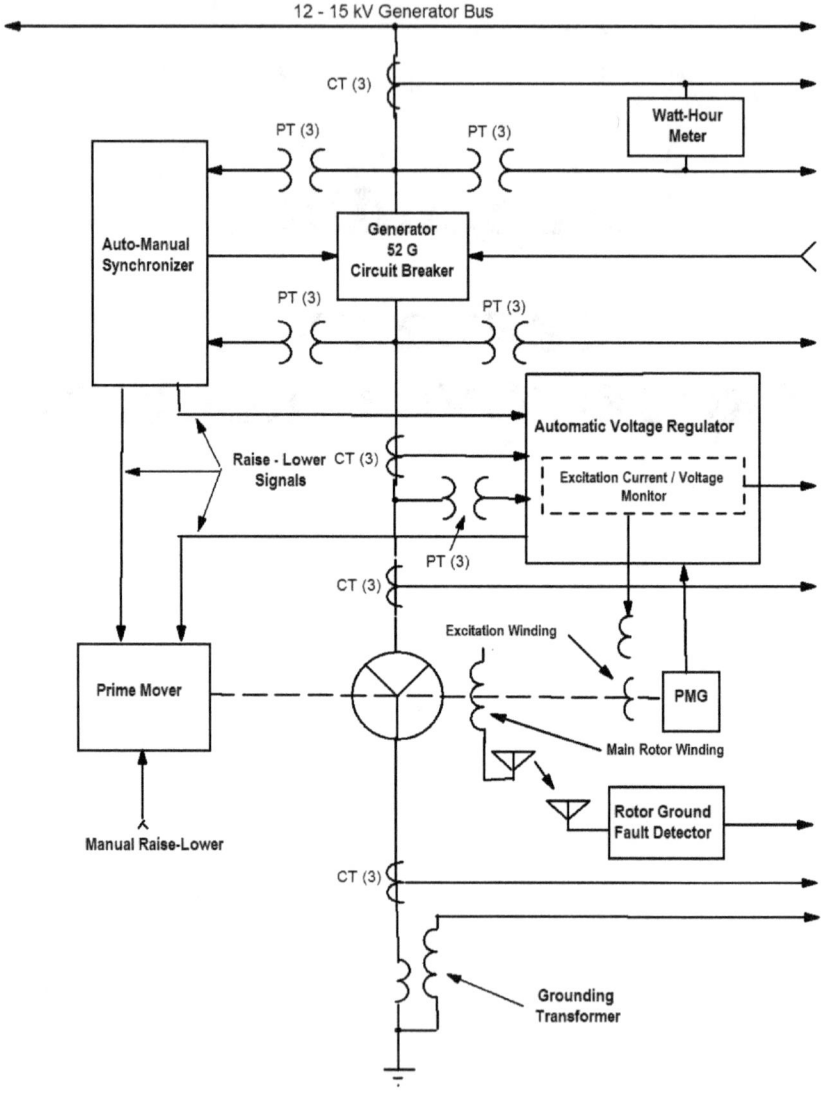

Generator Protective Circuitry

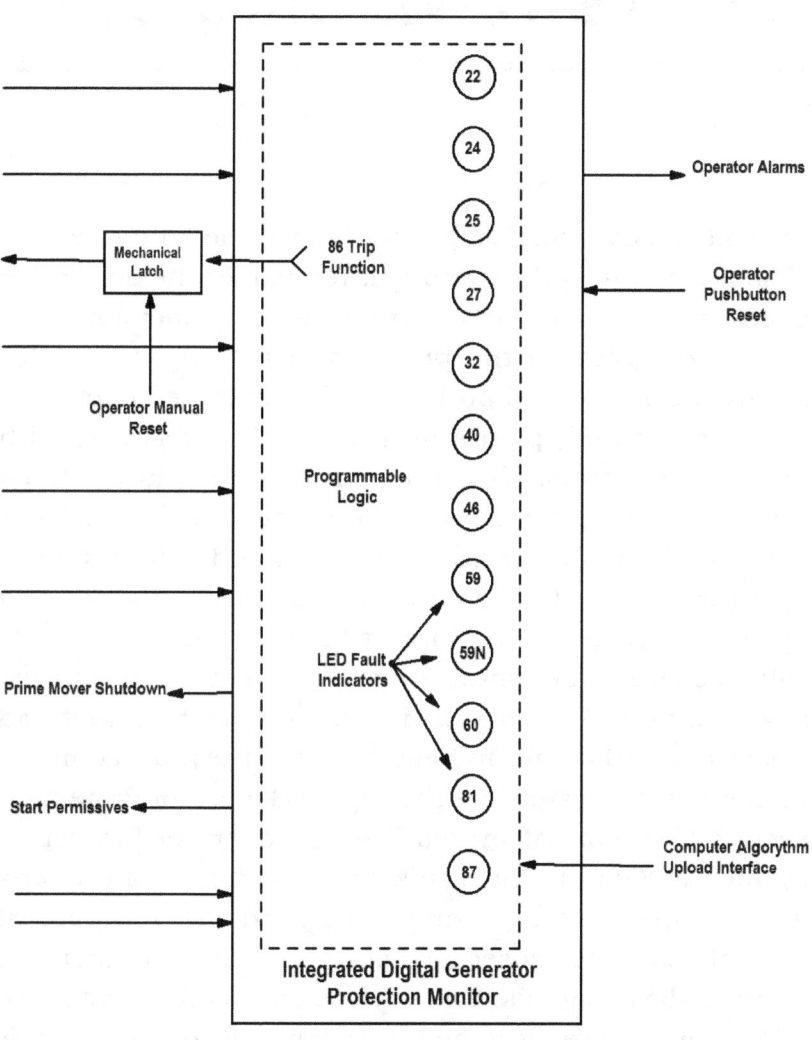

GENERATOR PROTECTIVE CIRCUITRY

This discussion will deal primarily with the fundamentals of safety fusing and switching intended for the protection of equipment within an electric generating facility.

The diagram on the previous facing pages provides an overview of typical protective circuitry used in electric power generating plants and how it is integrated with other subsystems. The protective system is centered around a programmable microprocessor-based unit designed to monitor critical currents and voltages, alert operators when damaging conditions are detected, and open the 52G generator circuit breaker if the monitored values exceed safe limits. Typically, the unit is mounted in a standard 19-inch-wide instrumentation cabinet and is less than 10 inches in height. Two of the most common names for this system are the Digital Generator Protection system (DGP) or Integrated Digital Generator Protection system (IDGP). In past history, monitoring functions were implemented using independent mechanical assemblies that closed a set of internal contacts to provide alarm signals. The devices were called protective relays and were assigned numbers designating their functions. A list of the function numbers as assigned by the American National Safety Institute (ANSI) is included at the end of this section. These numbered

Generator Protective Circuitry

functions remain in common usage; however, in modern systems they are digitally implemented within the IDGP software allowing adjustment as necessary to meet local conditions. For example, in some cases a slow rise in current can be permitted until a specified time limit is exceeded before opening the 52G circuit breaker. At other monitoring points, different rates of change may be required, or in some cases, the difference between two currents or voltages may signal a fault. Rate-of-change, voltage and current values can be uploaded through a communications port on the IDGP. Furthermore, when a fault occurs, the sequence of events that caused the 52G generator trip signal are recorded within the IDGP and can be retrieved to re-construct the primary cause of the fault.

The IDGP panel provides Light Emitting Diodes (LEDs) to indicate fault conditions. Any single fault energizes its respective LED and generates a signal that latches the 86 function. The 86 signal activates a trip signal to the 52G circuit breaker switch gear and latches into that condition. The 86 latch is manually reset by operators but is inhibited from reset until the condition or conditions that caused the 52G trip are cleared and an LED reset pushbutton on the IDGP panel has been activated. Operators are cautioned not to reset the fault indication circuitry unless the reason/s for the fault is thoroughly understood.

The IDGP monitor also provides signals to shutdown or throttle back the prime mover when appropriate and prohibit restarting the prime mover until faults are cleared.

A current transformer is installed at each end of

the three generator stator windings. The 87 protective function, measures the difference in current flow into and out-of each stator winding. Should the two currents differ more than a pre designated value, the 87 signal activates to trip the 52G circuit breaker. A difference in input and output stator currents indicates current is flowing to ground or to another winding within the generator requiring shutdown, troubleshooting, and repair within the generator. The transformer in the common grounding leg of the generator is designated a grounding transformer "GT" whose secondary feeds the 59N protective function. The output of the grounding transformer should always be near zero because at every instant in time the algebraic sum of all three generator phase currents is zero. An output other than zero indicates a ground fault within the generator or a significant phase imbalance in the current demand from an unbalanced load.

A list of ANSI safety device numbers is included below for easy reference when discussing safety functions.

In the design of electrical power systems, the **ANSI Standard Device Numbers** (ANSI /IEEE Standard C37.2) denote what features a protective device supports (such as a relay or circuit breaker). These types of devices protect electrical systems and components from damage when an unwanted event occurs, such as an electrical fault. Device numbers are used to identify the functions of devices shown on a schematic diagram. Function descriptions are given in the standard.

ANSI/IEEE C37.2-2008 is one of a continuing series of revisions of the standard, which originated in 1928.

Generator Protective Circuitry

LIST OF DEVICE NUMBERS AND ACRONYMS

- 1 – Master Element
- 2 – Time Delay Starting or Closing Relay
- 3 – Checking or Interlocking Relay
- 4 – Master Contactor
- 5 – Stopping Device
- 6 – Starting Circuit Breaker
- 7 – Rate of Change Relay
- 8 – Control Power Disconnecting Device
- 9 – Reversing Device
- 10 – Unit Sequence Switch
- 11 – Multi-function Device
- 12 – Overspeed Device
- 13 – Synchronous-speed Device
- 14 – Underspeed Device
- 15 – Speed – or Frequency, Matching Device
- 16 – Data Communications Device
- 17 – Shunting or Discharge Switch
- 18 – Accelerating or Decelerating Device

Electric Power System Fundamentals

- 19 – Starting to Running Transition Contactor
- 20 – Electrically Operated Valve
- 21 – <u>Distance Relay</u>
- 22 – Equalizer Circuit Breaker
- 23 – Temperature Control Device
- 24 – Volts Per Hertz Relay
- 25 – Synchronizing or Synchronism-Check Device
- 26 – Apparatus Thermal Device
- 27 – Undervoltage Relay
- 28 – <u>Flame detector</u>
- 29 – Isolating Contactor or Switch
- 30 – <u>Annunciator</u> Relay
- 31 – Separate Excitation Device
- 32 – Directional Power Relay
- 33 – Position Switch
- 34 – Master Sequence Device
- 35 – Brush-Operating or Slip-Ring Short-Circuiting Device
- 36 – Polarity or Polarizing Voltage Devices

Generator Protective Circuitry

- 37 – Undercurrent or Underpower Relay
- 38 – Bearing Protective Device
- 39 – Mechanical Condition Monitor
- 40 – Field (over/under excitation) Relay
- 41 – Field Circuit Breaker
- 42 – Running Circuit Breaker
- 43 – Manual Transfer or Selector Device
- 44 – Unit Sequence Starting Relay
- 45 – Abnormal Atmospheric Condition Monitor
- 46 – Reverse-phase or Phase-Balance Current Relay
- 47 – Phase-Sequence or Phase-Balance Voltage Relay
- 48 – Incomplete Sequence Relay
- 49 – Machine or Transformer, Thermal Relay
- 50 – Instantaneous Overcurrent Relay
- 51 – AC Inverse Time Overcurrent Relay
- 52 – AC <u>Circuit Breaker</u>
- 53 – Exciter or DC Generator Relay
- 54 – Turning Gear Engaging Device

Electric Power System Fundamentals

- 55 – Power Factor Relay
- 56 – Field Application Relay
- 57 – Short-Circuiting or Grounding Device
- 58 – Rectification Failure Relay
- 59 – Overvoltage Relay
- 60 – Voltage or Current Balance Relay
- 61 – Density Switch or Sensor
- 62 – Time-Delay Stopping or Opening Relay
- 63 – Pressure Switch
- 64 – Ground Detector Relay
- 65 – Governor
- 66 – Notching or Jogging Device
- 67 – AC Directional Overcurrent Relay
- 68 – Blocking or "Out-of-Step" Relay
- 69 – Permissive Control Device
- 70 – Rheostat
- 71 – Liquid Level Switch
- 72 – DC Circuit Breaker
- 73 – Load-Resistor Contactor

Generator Protective Circuitry

- 74 – Alarm Relay
- 75 – Position Changing Mechanism
- 76 – DC Overcurrent Relay
- 77 – Telemetering Device
- 78 – Phase-Angle Measuring Relay
- 79 – AC Reclosing Relay
- 80 – Flow Switch
- 81 – Frequency Relay
- 82 – DC Reclosing Relay
- 83 – Automatic Selective Control or Transfer Relay
- 84 – Operating Mechanism
- 85 – Communications, Carrier or Pilot-Wire Relay
- 86 – Lockout Relay
- 87 – Differential Protective Relay
- 88 – Auxiliary Motor or Motor Generator
- 89 – Line Switch
- 90 – Regulating Device
- 91 – Voltage Directional Relay

Electric Power System Fundamentals

- 92 – Voltage and Power Directional Relay

- 93 – Field Changing Contactor

- 94 – Tripping or Trip-Free Relay

- 95 – *For specific applications where other numbers are not suitable*

- 96 – *For specific applications where other numbers are not suitable*

- 97 – *For specific applications where other numbers are not suitable*

- 98 – *For specific applications where other numbers are not suitable*

- 99 – *For specific applications where other numbers are not suitable*

SUFFIXES AND PREFIXES

A suffix letter or number may be used with the device number; for example, the suffix N is used if the device is connected to a Neutral wire such as 59N when the relay is used for protection against Neutral Displacement. Suffixes X, Y, Z are used for auxiliary devices. Similarly, the "G" suffix denotes a "ground", hence a "51G" is a time overcurrent ground relay. Suffix numbers are used to distinguish multiple "same" devices in the same equipment such as 51-1, 51–2.

Device numbers may be combined if the device

Generator Protective Circuitry

provides multiple functions, such as the instantaneous/time-delay AC over current relay denoted as 50/51.

For device 16, the suffix letters further define the device: the first suffix letter is 'S' for serial or 'E' for Ethernet. The subsequent letters are: 'C' security processing function (e.g. VPN, encryption), 'F' firewall or message filter, 'M' network managed function, 'R' router, 'S' switch and 'T' telephone component. Thus a managed Ethernet switch would be 16ESM.

Apendix D

GLOSSARY OF TERMS

Glossary of Terms

The definition of terms in this glossary apply to their use in electric power systems and are not necessarily accurate for other applications.

A

AC Electric Power – AC Power = VA(Cos θ) where, θ = displacement angle between Voltage (V) and current in Amperes (A).

Air-cooled – Cooling provided by air flow.

Alternating Current – Current that alternates from one value to another. In U.S. power systems, current alternates in a periodic sinusoid at the rate of 60 cycles per second.

Ampere – Unit of current flow named for French scientist Andre-Marie Ampere.

Amplitude – Difference between maximum and minimum current or voltage values or the difference between the maximum and minimum values of any wave form.

Angular Difference – The difference between two specified angles.

Anode – The terminal of a two terminal electrical device into which conventional current flows.

Electric Power System Fundamentals

ANSI – American National Standards Institute.

Automatic Protection Circuitry – Circuitry which performs protective actions automatically without operator involvement.

B

Base Load – The minimum load that must be served continuously in an electric power system.

Battery – A device consisting of one or more electrochemical cells that produce electron flow through an external circuit.

Blackout – A condition in which power is lost in all or in part of an electric power system.

Brush-Type Generator – A generator whose rotor excitation current is applied through brushes in contact with slip rings attached to its rotor.

Brushes – Mechanical contacts that "brush" against the moving rotor of a generator for the purpose of transferring current from the stationary to the rotating parts of the machine.

Brushless Generator – A generator whose rotor excitation current is supplied through magnetic fields rather than through brushes.

Bus – An electrical bus is a connection point along which the voltage is constant in both magnitude and phase angle.

C

Capacitive – The property of an electrical circuit that

Glossary of Terms

consists of more capacitive reactance than inductive reactance.

Capacitive Reactance – The property of an electric circuit that restricts current flow caused by capacitance. The unit of capacitive reactance (X_c) is defined as: $X_c = 1/2\pi fC$, where, f = frequency of current change in cycles per second and C is the capacitance of the circuit in Farads.

Capacitor – A capacitor consists of two metallic plates separated by a dielectric insulating material. In ordinary engineering practice, a capacitor takes the form of sheets of metal separated by insulating material rolled into a cylindrical shape or a flatter shape for very small capacitors.

Cascading Failure – A self-propagating failure.

Cathode – The terminal of a two terminal electrical device from which conventional current flows.

Centre-Tapped – With reference to a transformer, a centre tap is a connection at the centre of a winding allowing the induced voltage in an entire winding to be halved with respect to each end of the winding.

Circuit – A continuous loop through which electric current flows.

Circuit Breaker – An automatically or manually operated electrical switch designed to protect electrical circuits from damage caused by overloads or short circuit conditions.

Circuit Breaker Panel – A mechanical unit or package containing electrical circuit breakers.

Electric Power System Fundamentals

Coils – Electrical conductors wound in the form of coils.

Combustion Gas Turbine – An engine that converts expanding fuel-heated air into shaft horsepower.

Common Grounding Point – A point in an electrical circuit that is connected to earth ground.

Comparator – An electrical device that compares two or more signals and produces an output that is proportional to the difference between them.

Condenser – A device that stores static electricity is called a condenser or a capacitor. See capacitor above.

Conductor – An type of material which permits the flow of electric charges.

Contingencies – A series of events that could possibly happen.

Contingency Analysis – The prediction of results that could occur should one or more events take place.

Conventional Current Flow – In electric circuits, conventional current flows in the positive-to-negative direction. Electron flow (not conventional current) flows in the negative to positive direction.

Cooling or Heating Coils – Coils of tubing that carry liquid or gaseous compounds that remove or add heat to a confined environmental area.

Corona – Corona is the name given to visible arcing between two points of different voltage levels. Arcing across high voltage transmission line insulators is called corona and represents undesired current loss.

Glossary of Terms

Cosine – In trigonometry, the ratio of the length of the side of a right triangle adjacent to either of the non-90 degree interior angles divided by the length of the hypotenuse.

Cps or CPS – Cycles per second.

Commutator – A rotary electrical switch that permits reversing a generator's rotor coil connection to external circuits providing one-way current flow or direct current.

CT – Current transformer abbreviated (CT) used to measure currents flowing in AC circuits.

D

Daily Periodic Variation – The intermediate variation of load in the daily history of electrical loads between the minimum continuous load that must be served and the average peak load values that must be served in an electric power system.

Destabilizing Contingencies – Contingencies that would cause a power system to begin to shut-down for its own protection.

DGP – Digital Generator Protection systems are comprised of programmable microprocessor-based logic controllers that allow users to establish pre-defined limits for generating alarm or shut-down signals to prevent damage to expensive equipment such as generators, power transformers and power lines.

Dielectric – An insulating component in electrical devices.

Diode – A two terminal device that permits current flow in only one direction.

Diode Wheel – A mechanical assembly in a generator that provides mounting for diodes that rotate with the generator shaft.

Direct Current – An electric current that flows in only one direction.

Distribution Lines – Electric lines that carry electric power to customers in an electric power system. The generally accepted division between transmission and distribution level voltages is 100 kV. Distribution voltage levels in common use are 69 kV, 39 kV and 10 – 15 kV.

E

Economic Dispatch – The assignment of generator loads throughout a power system to achieve optimum cost.

Edison – American Inventor and first power plant owner who sold electric power commercially in the United States.

Electric Generator – A machine or device that converts energy into electricity.

Electric Magnet – A magnet with the North and South Pole properties of a permanent magnet but whose magnetic force is caused by electric current flow.

Electric Power – A quantity of power derived from an electric current flowing through an electrical load the unit for which is the watt.

Electrical Bus – An electrical connection point along which the voltage is constant in both magnitude and phase angle.

Glossary of Terms

Electrical Circuit – A loop through which current leaves a positively charged point and returns to a less positive or negatively charged point.

Electrical Load – Components of an electric circuit that absorb power.

Electricity – The set of physical phenomena associated with the presence and flow of electric charges.

Electromotive Force (EMF) – A force of attraction or repulsion that moves electrons.

Electron Current Flow – The direction that electrons are transferred through electrical conductors from negatively charged points to positively charged points. The direction of conventional current flow is from positively charged points to negatively charged points.

Energy – In physics, energy is an indirectly observed quantity which exhibits itself in many forms but has no specific definition. Einstein expressed energy as a quantity equal to mass times the velocity of light squared.

Engine Throttle – A mechanism that adjusts the speed and/or power output from an engine.

Excitation Current – As used in this text, excitation current is that current which flows through the rotating windings of a generator for the purpose of creating a moving magnetic field.

Exciter – The components of a generator that create excitation current.

Exciter Field Windings – In brushless exciters, exciter field windings are those windings that are stationary and receive direct current. The magnetic field thus created induces AC current into the rotating components of a generator exciter.

Exciter Rotor – The components of a generator exciter that rotate on the generator shaft.

Exciter Stator – The components of a generator exciter that remain stationary.

F

Faraday – English scientist who demonstrated for the first time that an electric current can be generated from a magnetic field. He also invented the induction coil and the electric transformer and explained electro-magnetic induction.

Fault – In electric power systems a fault refers to an electrical short circuit or current overload.

Fissionable Material – In nuclear physics and nuclear chemistry, nuclear fission is either a nuclear reaction or a radioactive decay process in which the nucleus of an atom splits into smaller parts (lighter nuclei), often producing free neutrons and protons (in the form of gamma rays), and releasing a very large amount of energy.

Flux – Lines of magnetic force

Foot-Pound – Units in pounds multiplied by distances in feet, i.e., force times distance.

Frequency – In electric power systems the number of alternating current cycles per second is expressed as its frequency.

Glossary of Terms

Fuse – A length of fusible wire that opens when heated by current flow. The value of current sufficient to "burn" and thus open the fusible wire is a calibrated value.

G

Galvani – An Italian scientist who studied muscular reactions to electric charges and is credited with significant contributions to the beginning studies of electricity.

Galvanize – To coat a metal, usually iron or steel, with zinc to prevent corrosion.

Galvanometer – A sensitive current measuring device.

Generation Efficiency – The mechanical power input to a generator compared to its electrical power output expressed as a percentage.

Generation Station – An electric power generating location.

Generator Synchronization – The procedure used to assure a generator's output matches the voltage, phase and frequency of an already powered system to which it will be connected before actually making the connection.

Grid – An interconnected network of generators and loads. An infinite grid is generally considered one in which the power contributed by a single generator is not greater than $1/20^{th}$ the total power supplied to the network.

Ground Fault – A fault in which a path for current flow has developed allowing an undesirable current to flow from the fault point to earth ground.

Grounding Transformer – A transformer installed such that current flow to earth ground through its primary windings is transferred to its secondary windings permitting detection of the current.

H

Half-cycle – The 180 degree portion of a 360 degree full cycle.

Hertz (HZ) – The designation used to express cycles per second.

Horsepower – A unit of power or work per unit of time defined as 33,000 Ft.-Lbs. per minute.

Hydro-electric – The generation of electricity from energy converted from falling water.

I

I^2R Losses – Expression for electric power lost caused by resistance.

IDGP – An acronym for Integrated Digital Generator Protection. See DGP above.

Impedance – The property of AC circuits equivalent to resistance in DC circuits.

Induced voltage – Voltage conveyed by induction through magnetic fields.

Induction – The production of a potential difference (voltage) across a conductor when it is exposed to a varying magnetic field.

Glossary of Terms

Inductive – The property of an electric circuit that causes current to follow an applied voltage.

Inductive Reactance – The property of an electric circuit that restricts current flow caused by inductance designated as X_L defined as: $X_L = 2\pi fL$, where, f = frequency of current change in cycles per second and L is inductance in Henrys.

Industrial Loads – A classification of electric loads that are caused by industrial users.

Inherent Losses – Losses that occur due to the normal characteristics of an electric load.

In-Phase – Voltage or current waveforms that occur simultaneously.

Instrumentation Transformer – A single transformer used to reduce voltages or currents to safe levels for measurement.

Instrumentation Cabinet – A housing in which electrical instrumentation is installed.

Intermediate Load – The historic daily periodic variation of electrical loads in an electric power system.

Internal Resistance – The resistance within a device or circuit element that exists regardless of the electrical loads that connected to the circuit element or device externally.

Iron Core – In transformers the material about which its windings are wound is called the core of the transformer. An iron core transformer would describe a transformer whose core is comprised of iron.

Island Mode – A generation mode in which a single generator or power plant is supplying loads independent of other generators.

Isochronous – The mode of generator operation described above, i.e., Island Mode.

J

Joule – A unit of power equal to one watt per second; also, 1.356 joules is equal to 1 ft-lb/sec. a quantity useful in translating electric power to mechanical horsepower, i.e. 746 watts equals 1 horsepower.

K

Kinetic Energy – Energy in the process of release such as falling water.

L

Laminations – Layers of metal bound together in transformer cores to minimize losses.

Latch – A device that remains in one state or position until reset to its opposite state or position. In general, a latch has a normal stable state and when changed can be reset to its normal state or condition.

LED – Acronym for Light Emitting Diode.

Line Impedance – The AC resistance of a power line or the combined resistive, capacitive, and inductive reactance of a power line.

Line Outage – An open transmission or distribution line.

Glossary of Terms

Line-to-Line Voltage – The difference in electrical potential between any two lines of a three-phase power line.

Liquid-Cooled – A cooling system in which heat is removed by the flow of a cooled liquid. A typical liquid cooling system uses coils of metal tubing through which a liquid coolant is circulated to remove heat.

Load Characteristics – The characteristics of electrical loads as capacitive, inductive, or resistive.

Load Flow – The direction and magnitude of power flowing in an electrical grid.

Load Flow Study – The determination of the direction and magnitude of power flowing in a specific electrical grid.

Longitudinal Slots – Slots or openings in a rotating device that are parallel to the axis of rotation.

Losses – Elements within an electric system that consume power but do not contribute useful work.

Lube Oil – A petroleum product that lubricates mechanical components.

Lubrication – The process of lubricating and thus preventing wear and heat build-up within mechanical devices.

M

Magnetic Coupling – The transfer of voltage or current by magnetic induction rather than by mechanical contact.

Electric Power System Fundamentals

Magnetic Field – An invisible force surrounding magnets whether permanent magnets or electric magnets or conductors carrying electric currents.

Magnetic Field Strength – The density of magnetic flux lines per unit area.

Magnetic Flux – Lines of magnetic force.

Magnetic Retention – The property of magnets to retain their magnetic field strength.

Mass – A physical substance that has weight and occupies space.

Mechanical Energy – An observed quantity of energy that exhibits itself in mechanical form such as the torque in foot-pounds exerted on a rotating shaft.

Milliamperes – One one-thousanth of an ampere or 1×10^{-3} amperes.

Motor – An electro-mechanical device that converts electric energy into rotational shaft horsepower.

MVA – Millions of volt amperes or 1×10^6 volts × amperes.

MW – Millions of watts or 1×10^6 watts

N

Natural Gas – A mixture of carbon and hydrogen that developed from the fossil remains of ancient plants and animals buried in the earth.

Negative Charge – There exist in nature two types of electric charges; one designated positive and one negative. Positively charged substances are repelled

Glossary of Terms

from other positively charged substances, but attracted to negatively charged substances. Negatively charged substances are repelled from negative and attracted to positive charges.

Nickel-Chromium-molybdenum – Name of metal alloy used in the construction of generator rotors manufactured by specific manufacturers.

North Pole – The end of a magnet from which magnetic lines of force emanate.

Nuclear Energy – Energy that results from the action of nuclear fission, a process in which the nucleus of an atom splits into smaller parts (lighter nuclei), often producing free neutrons and protons (in the form of gamma rays), and releasing a very large amount of energy.

O

One-line Diagram – An electrical diagram in which multiple electrical conductors are symbolized with single rather than multiple lines.

Open Circuit – An electric circuit in which the supply or return path of electrons is open.

Operator – An individual who is responsible for performing operations.

Over-Excited – A phrase attributed to generator exciters when they are receiving current in excess of their normal requirements.

P

Parallel – Electrical components or circuit elements that are connected in parallel with one another rather than connected in series with one another.

Peak Area – That part of a load curve that lies above the daily historic average.

Peaking Unit – A generating unit or power plant that provides power only when peak loads require it.

Permeability – The property of a magnet to retain its magnetism over time.

Phase Angle – Angular difference between two voltage or current wave forms.

Phase Imbalance – A condition that arises when electrical loads are not evenly divided between phases in a poly-phase system.

Phase Rotation – The direction in which phase voltages occur. In a three-phase power system the phase voltages are designated (A), (B), and (C). Should they occur in reverse order, i.e., (C), (B), (A) their rotational order is reversed.

Photo Transistor – A transistor semi-conductor in which the "base" of the transistor is biased "ON" by light from an external source causing current flow through the transistor to increase or decrease.

PMG – An acronym for Permanent Magnet Generator

Pole Slip – A condition in which the rotor of a generator rotates faster than its output power frequency.

Pole-pair – A pair of North and South magnetic poles.

Poly-phase – A generation system in which more than

Glossary of Terms

one output is derived per each 360 degree rotation of a generator rotor.

Potential Transformer – A transformer with many more turns on its primary winding than on its secondary winding for the purpose of stepping-down high voltages for voltage measurement purposes with minimum burden on the high voltage line to which it is attached.

Power – Work in foot pounds per unit time such as in units of horsepower defined as 33,000 ft.-Lbs. per minute.

Power Factor – The trigonometric Cosine of the angle between voltage and current in a power system.

Power Interchange Transaction – A monetary transaction between two entities who periodically exchange power with one another.

Power Plant Efficiency – A measure of heat converted into electric power expressed in BTUs per kilowatt hour or British Thermal Units per thousand watts of generated electric power.

Power Transformer – A transformer designed to elevate or decrease power line voltages with minimum power loss.

Power Triangle – A graphical representation of Real Power, Volt-Amperes, and Reactive Volt-Amperes (VARS) constructed as a right triangle illustrating how any one of the three parameters may be calculated from the other two.

Pressure Oil Seal – A mechanical seal designed to prevent the flow of oil that uses air pressure to assist in the sealing mechanism.

Electric Power System Fundamentals

Primary Coil – In a transformer the primary coil is defined as the driven coil; the coil from which the output is taken is called the secondary coil.

Prime Mover – In electric power plants a prime mover is the mechanism that provides mechanical power necessary to rotate the shaft of a generator.

Programmable Microprocessor-based Device – An electrical assembly that uses a solid-state microprocessor to carry out pre programmed actions. These devices are sometimes called micro-computers.

Protective Relay – A device whose purpose is to initiate a protective action should pre-determined limits be violated.

PT – An abbreviation for a potential transformer

R

Reactor – That part of a nuclear power plant in which a nuclear reaction takes place for the purpose of generating heat.

Real Power – In A-C circuits, real power is calculated by multiplying the applied voltage and current and the Cosine of the angle between them.

Rectifier – An electronic device in which alternating current is converted to direct current.

Reserve – In electric power systems, reserve power is the power generating capacity in excess of the present usage level.

Residential Load – Electrical loads comprised of the

Glossary of Terms

type of loads generally required by residential users of electric power.

Resistance – That property of electric circuits that resists or opposes the flow of electric current. Resistance is defined by Ohm's law as $R = V/A$, where V = voltage is volts, A = current flow in amperes, and R = resistance in ohms.

Resistive – A load that is resistive in character, i.e., the component of loads that are in-phase with the applied voltage.

Rotor – That part of an electric generator that rotates and contains the rotor windings.

Rotor Ground Fault – A fault within the rotating windings of a generator causing current flow between the windings and earth ground.

S

Secondary Coil – In a transformer, the coil receiving driving current is designated the primary coil or winding and the other, the secondary coil or winding.

Set Point – In control circuitry a set point defines the desired operational point.

Shaft-mounted Fan – In air cooled generators, fan blades on the rotating shaft increase the flow of air flowing over and through the generator components for the purpose of cooling.

Single-pole rotor – In alternating current generators a single-pole rotor is one that has one North and one South magnetic pole.

Sinusoidal – A wave form that is sinusoidal in shape, i.e., in the shape of a Sine wave.

Slip rings – Mechanical rings that rotate on a generator shaft in contact with stationary "brushes" that carry excitation current to the generator rotor.

South Pole – That pole of a permanent or electro-magnet into which magnetic lines of flux flow.

Spinning Reserve – Electric generators that are spinning and ready to receive electrical loads. Rules vary on what constitutes spinning reserve. In some states power that can be brought on-line within 15 minutes is considered spinning reserve.

Stationary – A mechanical object or component that does not move.

Stator – The stationary windings of a generator that do not rotate.

Stator Core – The core material that supports the stationary windings of a generator.

Stator Winding – The coils of wire that form the stationary parts of a generator into which voltage is induced from the rotor of the machine and from which power is transmitted.

Substation – The name given to a location containing transformers, buses, instrumentation, and other components for the measurement and distribution of electric power.

Subsystem – A division of a system or system within a system.

Glossary of Terms

Summing Junction – An electronic circuit symbolized as a circle with an "X" through the centre. The output from a summing junction (c) is the sum its inputs at (a) and (b) such that c = a + b.

Swing Bus – An electrical bus from which other buses are referenced.

Synch Lamps – Indicator lamps connected between the phases of a generator output and those of an already powered line to which the generator is being connected. When the lamps are extinguished the voltage between the generator and the line to which it is being connected is at a minimum.

Synchronization – The process of connecting a generator to another generator or power ssystem that is already powered. The end result of synchronization is achieved when the frequency, amplitude and phase of the generator and the already powered circuitry is the same.

Synchronism Equipment – Equipment that assists operators in the process of synchronization or performs the function automatically.

Synchroscope – A display that provides a visual indication of the phase angle difference between a generator and the circuitry to which it is being connected.

System Instability – A system is said to be stable if a single outage will not cause further outages. An unstable system is one that can lead to a cascading failure and total loss of load.

Electric Power System Fundamentals

System Load Flow Diagram – A drawing used to illustrate power flow throughout a power system.

T

Tesla – Serbian electrical engineer, who invented polyphase generators, motors, and transformers leading to the AC power systems in use today.

Three-phase Current – The combined current flowing in a three-phase circuit..

Threshold Detector – An electronic circuit that produces an output signal when a pre-determined level of current or voltage is reached.

Torque – The measure of force times distance the units of which are expressed in foot-pounds.

Transformer – An electrical device that converts voltages from one level to another.

Transient – An electrical phenomenon that occurs rapidly.

Transmission Line – An electrical conductor or conductors through which power is transmitted.

Transmission Voltage – The level of voltage in a transmission line or system.

Turbine – A mechanical device that converts the flow of a liquid or gas to rotational shaft horsepower.

Turns Ratio – An expression of the number of coil turns on the primary as related to the number coil turns on the secondary of a transformer.

Glossary of Terms

U

Unbalanced Load – A condition in which the load imposed on a poly-phase electric system is not evenly distributed among the phases.

Unit Commitment – The determination of when specific electric generating stations or generators are scheduled to produce power.

Uranium – A radioactive chemical element used as fuel in nuclear power plants.

V

Vector – An arrow that represents a quantity that has both direction and magnitude.

Volta – Italian physicist Alesandro Volta whose name has been given to the unit of electro-motive force, the volt.

Voltaic Cell – A battery.

W

Watt – The unit of electric power defined as one volt times one ampere.

Waveform – The shape of a periodically changing electric voltage or current algebraically plotted on a graph.

Weekly Load Curve – The historic normal loads for each day and hour of a week.

Westinghouse – American engineer and industrialist who provided funds in support of Nikola Tesla's experimental poly-phase electric power system.

Wind Turbines – Convert wind energy into rotational shaft horsepower

Z

Z (Impedance) – The combined properties of inductive, capacitive and resistive electric load components.

Answers to Questions

Answers to Questions

Section 1

1. Current flow in a conductor can be generated from a magnetic field.

2. Volta, in duplicating Galvani's experiment with frog's legs, learned the true source of electric current was from dissimilar metals that were in contact with the frog's legs, not "animal electricity."

3. Volta's "battery" provided a continuous source of electric enabling Oersted's accidental discovery that a relationship existed between magnetism and electric current.

4. If an electric current can generate a magnetic field, can a magnetic field generate an electric current?

5. Alternating current could be transmitted over greater distances with higher efficiency than direct current.

6. Alternating voltage could be raised and lowered using transformers, a phenomenon first demonstrated by the English scientist Michael Faraday.

7. Volta's invention was called a voltaic pile and was later called a battery.

Electric Power System Fundamentals

Section 2.1

1. Power plant efficiency is expressed as Btu per kWh.

2. Any three of the following four facts are suitable answers:

  a. An electric current is generated by relative motion between a magnetic field and an electrical conductor.

  b. When relative motion between the magnet and conductor is reversed, i.e., toward or away from each other, the direction of current flow also reverses.

  c. The magnitude or amplitude of the current produced by relative motion is directly proportional to the rate of change of motion.

  d. Current induced into a conductor is directly proportional to strength of the magnetic field through which the conductor is moved.

3. A mechanical pump moving liquid through a pipe is analogous to Faraday's experiment with a magnet and an electrical conductor as follows:

  a. Liquid in a pipe connected between the suction and discharge ports of a mechanical pump moves when the pump shaft is rotated just as current does in the electrical conductor when relative motion occurs between the magnet and conductor.

  b. Mechanical work must be exerted to move liquid flow or electric current flow.

  c. Current flow in the electrical circuit is caused by the electro motive force or voltage that is generated by the force of relative motion just

Answers to Questions

as the pressure difference between the suction and discharge ports in the mechanical analogy results in fluid flow in the pump circuit.

    d. When the valve in the pump liquid circuit is closed, the pressure difference between the suction and discharge ports remains present just as the voltage difference remains when the switch is open in the electrical circuit yet in both cases the flow of liquid or current is stopped.

4. Direct current flows in one direction only; alternating current flows into and out of its source in repetitive cycles.

5. The magnetic field surrounding conductors adds-to or reinforces the magnetic field surrounding adjacent turns in a coil such that the strength of the magnetic field surrounding a single conductor is multiplied by the number of turns in a coil.

6. Stator windings near both North and South rotor poles add the voltage induced by both poles therefore doubling the voltage induced from a single pole.

7. Brushless exciters avoid the mechanical wear associated with metal brushes in contact with moving slip rings on a generator rotor. Brushless exciters are not as responsive when rapid exciter current changes are required.

8. Diodes convert AC current induced from the stationary windings of a generator brushless exciter into direct current required for application to the rotor windings of a generator.

9. A prime mover provides rotational shaft horsepower to rotate the shaft of a generator.

Electric Power System Fundamentals

10. The shaft speed is calculated as follows:

11. 50 cycles per second × 60 seconds per minute or 3000 cycles (revolutions) per minute.

12. Increase generator cooling or reduce generator load.

13. The force of attraction or repulsion that moves electrons.

Section 2.2

1. Generator external controlling mechanisms are:

   a. Regulating the speed or power applied to the generator rotor from a prime mover, and

   b. Regulating the rotor excitation current.

2. Joule rotated a paddle wheel in a tank of water and measured the rise in temperature of the water that resulted from the work exerted in rotating the paddle wheel.

3. Generator efficiency is expressed as:

$$\text{Generator Efficiency} = \frac{\text{Mechanical Power Input}}{\text{Electrical Power Output}}$$

$$= \frac{\text{Driving Horsepower}}{\text{Watts Generated}}$$

$$= \frac{\text{HP}}{\text{W}}$$

4. Impedance is the name given to inductive, capacitive, and resistive elements when they are combined in electrical circuits. Reactance is the capacitive or inductive components of loads that impede the flow of current.

Answers to Questions

5. Capacitive reactance leads applied voltage.

6. Inductive reactance lags applied voltage.

7. Impedance (Z) is minimized when XL is equal to XC.

8. When VARS are positive, is capacitive.

9. An operator can increase rotor excitation current when the generator load is capacitive resulting in an increase in real power output.

10. A generator's operating point is located at the tip of the VA vector on the generator capabilities diagram.

11. When connected to an infinite grid, a generator's power output is regulated by increasing or decreasing its prime mover throttle position.

Sections 3

1. The voltage considered the dividing line between transmission and distribution levels is 100,000 Volts or 100 kV.

2. System loading and the distance to loads are primary considerations that determine the selection of transmission and distribution voltage levels. Distances greater than 40-to-60 miles provide realistic economic trade-offs for using transmission voltages above 100 kV.

3. The ratio of primary and secondary turns in a transformer are directly proportional to the primary and secondary voltages in a transformer.

4. Current is lowered as voltages are increased in power lines to provide the same power levels. Power line losses are equal to the square of current, therefore,

Electric Power System Fundamentals

when voltage is doubled, current is halved as are losses.

5. Instrumentation transformers provide voltage and current measurement.

6. Potential transformers (PTs) measure voltage; VTs reduce or raise voltage levels.

7. One-line wiring diagrams show more detail but require more drawing area. Three-line drawings require less drawing space but show less detail.

8. To designate their electrical winding configurations.

9. Energy Management System level drawings depict very large networks whose details are not necessary at the EMS management level.

10. Center tapped transformers allow using their full line voltage from end-to-end or half of their line voltage from the center tap to either end as applications require.

11. Residential end-user power is measured at the point of entry of the power line to the customer with a Watt-hour meter.

12. Three functions provided by high voltage switches:

    a. To isolate sections of a power grid that may be disabled or out of service,

    b. To provide optional routing of power flow, and

    c. To prevent damage to other connected equipment when faults are detected.

Answers to Questions

Section 4

1. A shift in phase angle between the two voltages causes a voltage difference at each instant in time that produces current flow.

2. Power flow is measured using data provided from potential transformers but is calculated using the following expression:

$$P = \frac{V_1 V_2 \sin \theta}{Z_L} \approx \frac{V^2 (\sin \theta)}{Z_L}$$

where,
- P = Power in Watts
- V = Voltage in Volts
- A = Current in Amperes
- Z_L = Line impedance in Ohms
- θ = the offset angle in degrees

3. Net interchange is the sum of all power exchanges with a system and its ties with other systems or billing entities.

4. A system one-line drawing depicts electrical interconnectivity; a power flow drawing depicts power flow between generating stations and system buses at various physical locations.

5. Contingency analysis predicts system instability or shut-down conditions.

6. The total system load is the difference between net interchange and total generation.

7. Economic dispatch determines the generation levels from each power generating plant within a system for maximum system efficiency.

Electric Power System Fundamentals

8. Intermediate loads are the daily periodic variation. Peak loads exceed the daily average.

9. Economic dispatch strives for maximum generation efficiency; unit commitment strives for determining a practical schedule for starting and stopping generators.

Bibliography

Brown, LeMay, Bursten, and Murphy, *Chemistry, the Central Science*, New Jersey: Pearson Printice Hall, 11th Ed., 2009

Cardwell, Donald, *The Norton History of Technology*, New York and London: W.W Norton & Company, 1994

Cheney, Margaret, *Tesla: Man Out of Time*, Barnes & Noble, 1981

Corcoran, George F., M.S. and Henry R. Reed, Ph.D., *Introductory Electrical Engineering,* New York: John Wiley & Sons, Inc, 1st Ed., 1957

Duff, John R. and Milton Kaufman, *Alternating Current Fundamentals*, 2nd Ed., 1980

Felsing, William A. and George W. Watt, *General Chemistry*, New York: McGraw Hill Book Company, 1951

Fitzgerald, A.E., Charles Kingsley, Jr. and Alexander Kusko, *Electric Machinery*, New York: Mc Graw-Hill Book Company, 1952

Gamow, George, *Biography of Physics*, New York: Harper & Row, Publishers, 1961

Gibbs, J.B, *Transformer Principles and Practice*, New York, Toronto, London: McGraw-Hill Book Company, Inc.: 1950

Electric Power System Fundamentals

Hill, Phillip G., *Power Generation*, Cambridge, Massachusetts, and London, England: The MIT Press, 1977

Kosow, Irving L, Ph.D., *Electric Machinery and Transformers*, 1972

Marcus, John and Neil Sclater, *Electronics Dictionary*, New York: McGraw-Hill, Inc., 5th Ed., 1994

Meyer, Herbert W., *A History of Electricity and Magnetism*, Burndy Library, 1972

Pansini, Anthony J., E.E., P.E., *Basic Electrical Power Transmission*, Rochelle Park, New Jersey: 1975

Pansini, Anthony J., K.D. Smalling, *Guide to Electric Power Generation*, Rochelle Park, Liburn, Georgia: 1994

Pera, Marcello, *The Ambiguous Frog*, Princeton University Press, 1992

Power Technologies, Inc., *Power Systems Training Course*: 1984

Romer, Robert H., *Energy An Introduction to Physics*, San Francisco, California: W. H. Freeman and Company, 1976

Tang, K.Y., *Alternating-Current Circuits*, Pennsylvania: International Text Book Company, 2nd Ed., 1957

www.ingramcontent.com/pod-product-compliance
Lightning Source LLC
Chambersburg PA
CBHW050636300426
44112CB00012B/1819